SOCIAL TV

SOCIAL TV
Multi-Screen Content and Ephemeral Culture

Cory Barker

University Press of Mississippi / Jackson

The University Press of Mississippi is the scholarly publishing agency of the Mississippi Institutions of Higher Learning: Alcorn State University, Delta State University, Jackson State University, Mississippi State University, Mississippi University for Women, Mississippi Valley State University, University of Mississippi, and University of Southern Mississippi.

www.upress.state.ms.us

The University Press of Mississippi is a member of the Association of University Presses.

An earlier version of parts of Chapter 4 was previously published in *Television & New Media* 18.5 (2017): 441–58.

An earlier version of parts of Chapter 5 was previously published in *The Projector: A Journal of Film, Media, and Culture* 15.2 (2015): 73–112.

Copyright © 2022 by University Press of Mississippi
All rights reserved

First printing 2022

∞

Library of Congress Cataloging-in-Publication Data

Names: Barker, Cory, 1988– author.
Title: Social TV : multiscreen content and ephemeral culture / Cory Barker.
Description: Jackson : University Press of Mississippi, 2022. |
Includes bibliographical references and index.
Identifiers: LCCN 2022004661 (print) | LCCN 2022004662 (ebook) |
ISBN 9781496840929 (hardback) | ISBN 9781496840936 (trade paperback) |
ISBN 9781496840912 (epub) | ISBN 9781496840943 (epub) |
ISBN 9781496840967 (pdf) | ISBN 9781496840950 (pdf)
Subjects: LCSH: Social media and television. | Social media—Influence. |
Mass media—Social aspects.
Classification: LCC HM742 .B357 2022 (print) | LCC HM742 (ebook) |
DDC 302.23/1—dc23/eng/20220210
LC record available at https://lccn.loc.gov/2022004661
LC ebook record available at https://lccn.loc.gov/2022004662

British Library Cataloging-in-Publication Data available

Dedication
To Rachelle, Hutchison, and Piccadilly

CONTENTS

Acknowledgments ix
Introduction: Best Photo Ever 3
Chapter 1: From TGIF to #TGIT: Simulated Liveness and Flow
   in Shondaland 25
Chapter 2: Immerse Yourself Deeper: Building AMC's
   Multi-Screen Storyworld 58
Chapter 3: Rewarding Viewing: Check-Ins and Social Productivity . . . 88
Chapter 4: "Great Shows, Thanks to YOU": Fansourcing and
   Legitimation in Amazon's Pilot Season 116
Chapter 5: "It's What Connects Us": HBO and Platform
   Authenticity on Twitter 144
Conclusion: Everyday Ephemeral Content 176
Notes 189
Bibliography 231
Index 245

ACKNOWLEDGMENTS

I am greatly indebted to those who have guided and aided me through this project over the last decade as Social TV transitioned from hyped to discarded phenomenon.

Thank you to the mentors, fellow graduate students, and colleagues who helped me develop my ideas in classrooms, coffee shops, Twitter chats, freelance columns, and conference presentations: Barbara Klinger, Elizabeth Ellcessor, Joan Hawkins, Ted Striphas, David Waterman, Amy Gonzales, Mark Deuze, Bärbel Goebel Stolz, Myc Wiatrowski, Noel Kirkpatrick, Katherine Lind, Saul Kutnicki, Daniel Grinberg, Rebecca Butorac, Jennifer Lynn Jones, Myles McNutt, Austin Morris, Max Dawson, Christine Becker, Chuck Tryon, Karen Petruska, Andrew Zolides, Kaitlin Thomas, Tim Surette, Kerensa Cadenas, Oriana Schwindt, Tony Adams, and Grace Wang.

A major thank you to Laura Portwood-Stacer and the members of the Summer 2019 Book Proposal Accelerator, who helped evolve the project in a much needed way. My gratitude is also with the team at the University Press of Mississippi, my editor Emily Bandy, and the reviewers, all of whom offered great ideas in the later stages of my work.

And finally, the biggest thanks to my family: to my parents, John and Sherry, for their facilitation of my lifelong obsession with TV; to Piccadilly, for her S-tier companionship and enthusiasm for head-clearing walks; and to Rachelle, for her endless support and sharp feedback as I finished this in the midst of the COVID-19 pandemic and the birth of our first child, Hutchison.

SOCIAL TV

Introduction

BEST PHOTO EVER

During ABC's live telecast of the Eighty-sixth Academy Awards on March 2, 2014, host Ellen DeGeneres walked down the aisle and spoke to stars Liza Minnelli and Lupita Nyong'o before stopping next to Meryl Streep. DeGeneres had a question for Streep and the global audience. "You are nominated a record-breaking eighteen times, right? So, I thought we would try to break another record right now with the most retweets of a photo." As DeGeneres described her selfie plan, a bewildered Streep interjected that the person sitting behind her, Julia Roberts, should also be in the photo. Roberts mockingly placed bunny ears near Streep's head, but DeGeneres was undeterred. The host recruited other nearby stars—including Nyong'o, Bradley Cooper, Jennifer Lawrence, Brad Pitt, and Angelina Jolie—to squeeze in for the photo. Excited murmurs spread across the Dolby Theatre as DeGeneres and Cooper bickered over who would take the picture and how to hold the smartphone for maximum coverage. Cooper finally snapped the image. The audience cheered. Streep mumbled to DeGeneres, "I've never tweeted before!" Moments later, DeGeneres posted the photo to her Twitter account (@TheEllenShow) with the caption, "If only Bradley's arm was longer. Best photo ever. #oscars."

DeGeneres got her wish. Within thirty minutes, the tweet attracted more than 700,000 retweets and nearly 200,000 favorites (now known as likes). In less than twenty-four hours, the tweet received almost 2.7 million retweets and 1.4 million favorites, nearly quadrupling the previous record set by a 2012 tweet sent by then-US president Barack Obama. As DeGeneres later announced during the telecast, the enthusiastic response to the tweet even managed to disable Twitter for 20 minutes. The outage inspired Twitter's corporate account to post, "The envelope please . . . to @TheEllenShow—this is now the most retweeted tweet with over 1 million RTs. Congrats!" The ensuing news coverage of the Oscars focused as much on how DeGeneres and friends "broke" Twitter as it did the usual winners, losers, and red-carpet fashion.[1] The moment has pervaded popular culture memory. *Time* magazine selected the selfie as one of the 100 most influential photos ever.[2] *The Simpsons* parodied it. Subsequent award show hosts have similarly tried to "go viral" with antics that exploit the chaotic potential of live television.

Notwithstanding the enthusiasm of Streep or Cooper, DeGeneres's tweet was not just a fun moment between celebrity friends. It was also the result of the collaboration between Twitter, ABC, Oscar producers, and sponsors. Fred Graver, then-head of TV at Twitter, later revealed that his team worked with DeGeneres and the telecast's producers to generate unique moments that would spike engagement on the social platform.[3] DeGeneres floated the celebrity selfie idea during one of these brainstorming sessions, but Graver and producers were unsure if the host would follow through. Samsung, which committed $20 million in promotional time for the telecast, pushed for the idea, and for DeGeneres to use its Galaxy Note 3 device for the photo.[4] Though DeGeneres skillfully orchestrated the scene on the fly, the virality of the image generated significant attention for Twitter, ABC, and Samsung.[5]

The photo was hailed as yet another sign of the game-changing impact of social media on television. This narrative swept through the media and technology industries in the late 2000s. Across trade conferences, trade presses, and product introductions, industry leaders proposed that social platforms like Facebook, Twitter, and venture capital-backed start-ups would shift television viewing into a multi-screen participatory experience accentuating the immediacy of live television. To hear executives and promotional campaigns tell it, television audiences would contribute to real-time conversations, engage more sincerely with programs, and explore bonus content on social platforms or corporate-backed mobile apps.[6] This imagined intersection between television and social media came to be known as Social TV.

To a degree, the circumstances of the "Oscar selfie" show the influence of social media on television. Here were A-list actors, known for tight control over their star personas, cramped together to take a cell phone picture. The Academy Awards ceremony is Hollywood's most prestigious event and one of television's biggest live events, but the crucial moment of the telecast chased social media buzz. Though the photo happened in a live amphitheater on television, it was distributed to Twitter to create conversation and inspire memes among the live audience and broader social media public. Industry discourse played into this distribution, stressing how DeGeneres captured social records, broke Twitter, and helped usher in a new vision for product placement. In the eyes of the media industries, the Oscar selfie was far more legible as a *social media event* than it was a television moment.

However, the moment still happened live—with a small delay—on television. Though preplanned, the time between when DeGeneres explained

Ellen DeGeneres's 2014 Oscar tweet, which she called the "best photo ever."

her plan to Streep and when Cooper snapped the photo was an acutely televisual moment where it felt like anything could happen. Without DeGeneres's performative questioning of Streep, Twitter users would not have known of the host's desire to break a record. Without the show's live television audience (44 million in the US), the tweet likely would not have hit the record at all, let alone in a few hours.[7] Despite the discourse angling to prove that the Oscar selfie typified social media's transformative effect on an "old" technology and medium, the photo showed that the US television industry had already synthesized social media into its normal practices. DeGeneres's selfie shenanigans were strategic, part of a now-familiar plan to generate social media attention for the Oscars that could extend the impact of the live telecast from one screen (television) to many others (phones, tablets, and laptops). Amid the social media hoopla, key industrial strategies related to television survived.

This book reveals how the television and tech industries promised—but failed to deliver—a Social TV revolution in the 2010s. Hollywood and Silicon Valley prophesized a world where the throng of screens in US households would synchronize to a collective, participatory experience. Social TV was pitched as a way to resist the progressively fragmented and on-demand nature of television but, more important, as a way to bring audiences together in shared virtual spaces and reward them for their viewership. Although Twitter received much of the attention, Social TV also emerged via television networks' mobile apps and start-up platforms that offered bonus content and digital prizes related to beloved series.[8] But the fan-friendly discourses about Social TV's potential elided predictable corporate maneuvering to monetize multi-screen data in an anxious, transitional moment before streaming redefined television—again.

I use Social TV to describe the intersections between social platforms and television, including 1) the schemes used by the media industries that combine television and social content; 2) the web-enabled second screens (phones, tablets, laptops) where viewers accessed television and social content; 3) the platforms, apps, and websites where viewers, industry professionals, and brands interacted regarding this content; 4) the practices legitimized or marginalized by social platforms' incursion into the viewing environment; and 5) the conversational threads produced about television and social platforms, by those within the respective industries, press, and consumers. Early analyses of Social TV examined the emergent multi-screen viewing habits and focused on Twitter as a central site of "backchannel" conversation or new promotional measures.[9] But the speed at which Social TV fizzled as an industrial concern also influenced its diminished presence in academic research. To a degree, the malleability of Social TV that aided its early hype also led to its dissolution. This book reconstructs the history of Social TV throughout the 2010s to show that, even in failure, the phenomenon still impacts how people interact with television and social media in a world of Facebook original series, streaming bundles, and unlimited content options.

REMEDIATING TV AND LEGITIMIZING SOCIAL TV

At first, Social TV merely repackaged familiar Hollywood content, promotion, and audience measurement strategies in new data-rich environments—what Hye Jin Lee and Mark Andrejevic refer to as "digital enclosures."[10] For the television industry, Social TV served as a way to appear responsive to shifts in technology and the activity of ascendant online fan culture. The movement enabled industry veterans to reiterate the aura of live broadcasting, the programmed schedule, and brand identity, all bolstered by an inventory of advertisements, product placements, and cross-promotions. Liveness and flow—described by Raymond Williams as the movement from episode to ad break to next episode—have long been seen as essential to the phenomenological experience of watching television.[11] But the industry also constructs liveness and flow so that they appear natural to viewers. With on-demand technologies posing a continuous threat to the aura and economic value of live, preprogrammed television, Social TV initiatives aimed to salvage liveness and flow for the modern era as well as to marshal consumers' multitasked viewing across screens and social platforms. The products promoted liveness strengthened by the connectivity of viewers and a flow synthesizing attention between television screens to mobile devices, leading early analyses of Social TV to refer to the efforts as "connected viewing," "co-viewing," or "co-connected viewing."[12] These modified forms of liveness and flow contributed to an imagined viewing environment that was collective and uniform and yet totally personalized for each participant. In this regard, Social TV tried to harness the contradictions of an increasingly fragmented, individualized experience under the banner of collective, productive engagement.

For tech companies, Social TV served as an entry point to partnerships with legacy media conglomerates and their immense capital and star power. It also perpetuated the enduring Silicon Valley mythology about the democratizing impacts of technology. By the second dot-com wave of the 2010s, tech companies publicly embraced an ethos of "disruption" and a desire to "make the world a better place."[13] Start-ups of all sizes have pitched "better, cheaper, faster" ideologies as solutions to rigid and regressive industries from public transit to health care.[14] While the societal impact of companies like Uber or Airbnb is hotly contested, a consequence of the obsession with Silicon Valley in the post-Great Recession America is that tech terminology has trickled into non-tech corners of life. The mantra of disruption—and the promise that new technology will solve problems missed or created by old technology—is part of a history of technological utopianism weaponized by inventors and politicians dating back to the creation of electricity and the steam engine.[15] But the press also plays a critical role in circulating these techno-utopian promises. The industry blog *TechCrunch* even hosts a popular annual showcase contest for "revolutionary" start-ups called *Disrupt* that was famously parodied on HBO's *Silicon Valley* for its outsized messaging about technological transformation.[16]

The extension of familiar strategies into the social arena was part of what I call a *remediation-legitimation cycle*, where new technology is positioned to *remediate* an old medium like television into a more interactive, communal, and democratic entity, only to be subsumed into *legitimate* (e.g., typical, revenue-generating) practices. Jay Bolter and Richard Grusin theorize remediation as how media reform, recontextualize, and respond to other media, both older and newer.[17] Though television remediates other media—it borrows from radio, film, and theatrical performance, among other forms—it has also been targeted for upgrades by processes of remediation. For Bolter and Grusin, the "rhetoric of remediation" intensifies the stakes of a potential technological shift.[18] Lisa Gitelman and William Boddy respectively argue that these "negotiations" and "public rehearsals" play out discursively, in promotional materials, magazine trend pieces, and legal standoffs.[19] As corporations promoted the disruptive potential of Social TV, they also exploited the imprecise import of activity and information generated within the ecosystem, distilled into metrics like engagement and buzz, to hide their technical function.

According to danah boyd, "social" implies a place for communities to "gather, communicate, share, and in some cases, collaborate or play."[20] Human-oriented discourse manifests as simply as the labeling of connections as friends, but it also arrives through discourse about posts attracting *social engagement* or calls for consumers to *join the conversation* using a hashtag. Though we might know that "engagement" equals a mix of likes, shares, and responses, there is no agreement as to what the actual significance of these forms of engagement is. We know that the "conversation" is what people are talking about on a platform, but users do not always speak directly to each other in the form of a discussion. This positivist, vague language valorizes consumer participation as well as disguises the technical function of platforms that complete tremendous amounts of data collection.[21] Thus, trade press reports and promotional campaigns espousing Social TV's power tried to convince industry constituencies—executives, sponsors, audience measurement companies like Nielsen—of this power as much as it aimed to convince consumers. This process normalized certain practices like live-tweeting new episodes that could be collected into ratings data and eventually monetized. Social TV products that could not be as easily integrated into industry narratives about collective fan conversations were deemed failed experiments, and in some instances, wiped from the web.

This remediative refrain has been prevalent throughout television's lifespan with new technology claiming to improve its flaws and to make it more educational, personalized, communal, or democratic. Cable was hailed in the 1960s as a way to evolve television beyond entertainment and into the realm of public service with continuing education programs, home banking, and virtual civic participation.[22] The mid-century furor over advertising's negative impact primed Americans to embrace rhetoric from manufacturers like Zenith that swore remote controls would "zap annoying programs and objectionable commercials."[23] VCR makers sold another romanticized vision of consumer control, celebrating the technology's role in the rise of connoisseurship and time-shifting embodied in the still-common guarantee to "Watch Whatever Whenever."[24] The proposed solutions of the mid-century gave way to more hyperbolic digital visions fueled by the 1990s obsession with interactivity. Futurists like Nicholas Negroponte seized on television's reputation as "the idiot box" by proposing that computing power would make it smarter.[25] George Gilder contrasted how television lulled viewers into a "stupor," while computers enabled users to become "richer and smarter and more productive."[26] Phillip Swan promised that computer-assisted televisions would become so smart that viewers would "have to be educated" about their capabilities.[27]

These accounts demonstrate the incongruities within idealistic visions for television and within remediation-legitimation cycles. The perception of cultural frustration with television in the mid-twentieth century produced the right conditions for utopian promises to emerge. While cable, the remote control, and the VCR caught on with consumers, the innovations did not "fix" television, nor did they eliminate the nuisance of ads and sponsors. Indeed, their popularity only intensified US television's relationship with targeted ads and product placement. As each new invention failed to prevent television from being overrun by ads and distasteful programming, another arrived to save the day.

Meanwhile, the television industry has viewed disruption skeptically to continue delivering live audiences to sponsors. The resistance manifested through disciplining of consumer behavior and criticism of industry partners. As 1980s viewers zipped through channels and commercials, network leaders questioned their ability to deliver accurate information in Nielsen diaries.[28] When Nielsen later found that the diaries underestimated the cable audience up to 45 percent, networks called for new "correct measurement" practices.[29] By the mid-1980s, the industry knew that millions used VCRs, but conflicting reports could not determine how, when, or the impact on ad recall.[30] The conjecture festered paranoia that both the VCR and Nielsen needed to be "meticulously scrutinized" for their role in upsetting the "neat and ordered business" of ratings.[31] Pleas for revisions of the diaries and more VCR information led to the creation of Nielsen's "People Meter," which asked participants to chart their viewing via a device connected to the television.[32] But again, once the new system revealed lower ratings for broadcasters, network heads raised concern about viewers having to do "data entry."[33]

A similar pattern emerged with the computer-powered DVR, which was initially condemned for its commercial-skipping technology. After manufacturer TiVo released a rebellious ad campaign featuring television executives being tossed from the window of a boardroom, the industry went on the offensive.[34] Turner Broadcasting's Jamie Kellner equated fast-forwarding through ads with a DVR to stealing.[35] In 2001, more than twenty-five networks sued TiVo's competitor, ReplayTV, for an "unlawful scheme that attack[ed] the fundamental economic underpinnings of free television non-broadcast services."[36] Network leaders also blamed the "nervous" advertising industry for its resistance to change in the face of declining live viewership and complained about the failure of the Nielsen "monopoly" to evolve alongside modern technology.[37] But despite the public protests, networks and studios saw the DVR's potential to offer detailed information about viewers and subsequently pressured Nielsen to develop multiple audience tracking initiatives: first, a deal with TiVo and second, a full-scale "DVR rating" that tallied viewership up to seven days after an episode's original live airing.[38]

The turbulent response to new technology and its impact on live viewership demonstrates the contradictory positions of the television industry. For years, network executives both critiqued and empathized with consumers over their alleged disorderly behaviors. They supported Nielsen's methodology until new technology and viewing practices hinted that live viewership was on the decline. They demonized technology as troublesome and indoctrinating of unwitting consumers until they needed *other* new processes to better tabulate ratings that would not diminish ad rates. Altogether, the responses show the industry's desire to maintain existing partnerships and models at all costs. It is not that networks see all technological innovations as bad; it is only those that threaten the vital practices that are worth attacking, maligning, and urging partners to overcome. But once the utility of an insurgent technology is identified, it is legitimized within existing schemes and framed as beneficial for everyone. The contradiction of remediation is that new technology is both problem and solution.

Social TV emerged during another crisis moment for the US television industry. Networks spent the second half of the 2000s trying to survive the rise of the web, high-definition video streaming, and mobile accessibility by extracting value from what John T. Caldwell calls "second-shift" content via a litany of distribution methods.[39] They turned to mobisodes, character blogs, alternative reality games (ARGs), digital comics, podcasts, and transmedia storytelling tactics to sustain online interest in programs between episodes and seasons.⁴⁰ Studios embraced digital stores like iTunes and Amazon Video and streaming "portals" like Netflix and Hulu as evolutions in the home video market. Some executives even argued that ARGs, mobisodes, and video streams could inspire more passionate fans to watch live, ad-supported episodes.⁴¹ But between DVRs, digital stores, and streaming portals, live viewership for non-sports programming began to fall sharply by the early 2010s.⁴² Fears about "cord-cutters," consumers who ditched their cable packages, percolated in the trades and business sections of national publications.⁴³ Led by Apple's iPhone and iPad, the abundance of smartphones and tablets enabled more mobile and on-demand viewing. The anxieties about time- and place-shifting and even greater audience segmentation created the proper climate for Social TV to be viewed as another techno-savior.

Indeed, though early experiments with social media—including corporate-run Facebook fan pages and participatory ARGs—were met with resistance or dissonance, once networks toyed with forms of live engagement, Social TV was treated more seriously. Fox devised "tweet-peat" events that displayed a ticker of Twitter commentary from viewers, producers, and performers at the bottom of the screen during September 2009 reruns of *Fringe* and *Glee*.⁴⁴ The promotion drew criticism from viewers for the distracting placement of the ticker, but Fox executives were pleased to learn more about the audience's nascent multitasked, multi-screen habits.⁴⁵ NBC promoted the September 2010 premiere of *Community* with a "Twittersode," which saw accounts for the sitcom's characters exchange tweets in real time using an approved hashtag (#NBCCommunity) in the hour before the episode's debut.⁴⁶ By spring 2011, the use of hashtags to live-tweet new episodes or major events had become standard practice. Comedy Central then popularized the use of hashtags as on-screen chyrons after the *Roast of Donald Trump* (#trumproast) generated more than 27,000 tweets in a few hours.⁴⁷ A month later, CBS expanded live Twitter promotion with its "#TweetWeek" campaign, where the network's stars fielded questions directed to a hashtag (e.g., #Survivor) during live episodes.⁴⁸

Meanwhile, armed with a fast-growing user base and no clear revenue model, Twitter embraced its role as a collaborator before it could be pegged as an antagonistic disruptor. In a blog post, co-founder Biz Stone argued that Twitter's real-time integration into coverage of the 2008 US presidential election demonstrated "how Twitter could make television more interactive and possibly even have a democratizing effect on the medium."⁴⁹ Twitter regularly used its blog to disseminate data regarding increases in user activity during live televised events like the Super Bowl and award shows.⁵⁰ The company

#TrumpRoast was one of the first on-screen hashtag chyrons.

also published guides for networks and celebrities looking to craft engaging live-tweeting events.[51] Twitter later introduced Amplify, which allowed networks to purchase sponsored posts during live broadcasts and monetize them with pre-roll third-party ads.[52] Twitter declared that its integration into the television experience constituted "the new water cooler," with hashtags positioned as "the front door to the shared conversation."[53] Crucially, Twitter's pitch stressed the mutual benefit of the relationship; the company did not attack television for its failures but rather angled to improve the viewing experience for everyone.

Similarly, after facing staunch criticism from partners in prior eras, Nielsen tried to proactively measure social activity and its relationship to live viewership. A 2011 study of 150 million social posts and 250 television series discovered a "statistically significant relationship" between posting and watching, particularly among the 12–17 and 18–34 age groups.[54] Nielsen and partner NM Incite determined that weekly peaks in social posts about television correlated with viewership.[55] By 2012, a cascade of overlapping reports found that while watching television, 40 percent of viewers used a second screen, 50 percent chose the smartphone, and 60 percent used tablets.[56] Research firm Trendrr found an 800 percent growth in Twitter chatter about television from 2011 to 2012, with every month drawing at least double the number of posts compared to the prior year.[57] In response, Nielsen unveiled the Twitter TV Ratings, a metric for reporting the total number of tweets and their reach (or how many people saw them) related to television. Nielsen and Twitter partnered on the new rating to serve their Hollywood partners and maintain their viability in the Social TV space. Network executives declared that the data should be instantly viewed as legitimate because it would help all related industries.[58] The adherence to data—in ambiguous forms of reach, impressions, and check-ins—had the same purpose as Nielsen's television or DVR ratings: to convince sponsors that social platforms could facilitate lucrative multi-screen initiatives for industry content and to demonstrate to viewers a standardized system for how to participate in the Social TV ecosystem.

Viewed this way, Social TV never intended to revolutionize television as promised in promotional campaigns or press speculation. But it also failed to sustain consumer interest in live, synchronized, and collective viewing, particularly as Netflix increased its spending on licensed and original programming across the 2010s. Legacy media companies, previously invested in upholding the status of television as a preprogrammed and live product, have been forced to partner with Netflix and develop in-house streaming enterprises. Award show hosts continue to try to top the Oscar selfie with constructed chaos intended to trend on Twitter, but Social TV has fizzled as a dedicated industry project.

Nevertheless, the relationship between television and social media is now more potent than ever. Though networks struggle to persuade modern viewers to watch a new episode live, they have continued to exploit the perceived value of engagement to further embed their brands onto every screen, every platform, and every moment of life. Likewise, Twitter and Facebook used the initial Social TV wave as a springboard to develop and distribute television programs of their own. With television networks posting memes and hashtags on social platforms and social platforms distributing episodes of television, the boundaries between social and television have collapsed into an endless ecosystem of ephemeral content.

The shift in emphasis to creating endless content is important. It relies on television networks developing what Sarah Banet-Weiser calls "authentic" social brands that augment—but do not undermine—the preexisting brand identities related to television programs.[59] Media companies now micro-target consumers with multi-million-dollar franchises as well as with comedic Facebook posts, YouTube clips, and hashtag activism on Twitter. Rather than attempt to convert social media chatter into live television viewership, these corporations now look to generate consistent engagement with social media users at any time of the day. Thus, while the initial Social TV strategy failed to meet the early industry hype, this book uncovers that it succeeded as part of an unending corporate media project to control consumer attention on a global scale one swipe at a time. Ethan Tussey sees this shift as part of a "procrastination economy" focused on capturing our attention on mobile devices during "in-between moments."[60] To an extent, the procrastination economy extends from television's successful growth into domestic spaces and routines in the twentieth century.[61] But the modern focus on monopolizing attention is also more predicated on a hyper-charged version of network brand identity, which has been an essential tool of audience management since the cable expansion.[62] Social platforms and digital content drive a more sophisticated form of "symbolic" or "affective" relationship branding that attempts to *eventize* as many possible moments of everyday life.[63]

EPHEMERAL HISTORIES

Social TV was a fundamentally ephemeral phenomenon. As Paul Grainge explains in his introduction to a collection on ephemeral media, the term "describes anything short-lived."[64] Grainge argues that ephemerality "signifies the relation of media forms to *regimes of time* (duration, shortness, speed) and *regimes of transmission* (circulation, storage, value)."[65] By design, Social TV strategies aimed to structure viewer experience across both of these regimes. On-screen hashtags and second-screen apps tried to strengthen the temporal ephemerality of the televisual moment. They tried to convince people to tune in live lest they miss valuable content on multiple screens. But at the same time, the industry wanted to expand the circulation of social chatter and unbundled digital content.[66] Rather than challenge the supposed short attention spans and endless conversations occurring on social platforms, Social TV strategies harnessed the ephemerality as a modern feature of television programming. Individual tweets, status updates, likes, and clicks also represent what Steven Schneider and Kristen Foot call a "unique mixture of ephemeral and permanent."[67] They may not have significant import, but they still represent audience interest in ways that the media industries have deemed useful.

Beyond content strategy, the Social TV hype cycle reflects the ephemerality of big industry ideas and trends. Social TV was the future—until it was not. As Social TV struggled to reach market saturation, television and tech companies moved on. Mobile apps with lucrative investment deals with media conglomerates disappeared from the web and digital stores. Networks terminated their two-screen products with no fanfare, shifting them from hyped objects to what Amelie Hastie calls forgotten media "detritus."[68] Multiple Social TV projects conjured the same media coverage cycle of hype, investment, and, eventually, indifference. The products live on in collections of archived links, blog posts, screenshots, videos, and snippets of user activity. While, as Elizabeth Evans notes, the internet can be perceived as "anti-ephemeral," creating an endless circulation of content, Social TV shows how corporate digital minutia can just as quickly turn "hyper-ephemeral," disappearing in an instant.[69] The perceived lack of consumer interest subsequently justified a lack of coverage on pivots away from an ephemeral trend. We should not be surprised when products situated *as* ephemeral to inspire engagement consequently *become* ephemeral.

Social TV existed so briefly as an industry concern that is best described as a historical micro-moment. Situating this era as a micro-moment follows Helen Wheatley's claim that we can produce "dynamic interventions that question history and historiography of television more widely" by exploring specific key moments of change.[70] Thus, the study of Social TV requires what I call *ephemeral historiography*, or analysis of the once hyped but eventually failed, temporary, or discarded. This method is inspired by Rick Altman's "crisis historiography," which details how new media surface without definition and secure meaning and function through their incorporation—or disruption—of society's habits. Altman argues that crisis historiography assumes that "the definition of a representational technology is both historically and socially contingent."[71] That requires scholars to reconstruct the specific historical and social contexts in which technology emerges but also how individuals and groups negotiate the definitions of that technology's identity. Most relevant to Social TV, Altman stresses that crisis historiography also necessitates attention to the "technological dead ends and technical failures" that glowing histories of technological progress often minimize.[72] Ephemeral historiography places dead ends and failures at the center of its analysis, even more than the crisis framework. These brief "flashes" or "raids and spot checks" on history work to combat what Wheatley calls scholarship's tendency to "inherit" the media industries' inflated claims about the "next big thing."[73] Addressing failure reveals a greater insight of the circumstances from which new developments materialize and how stakeholders pivot to the next disruption. Tracing the history of ephemeral phenomena like Social TV from beginning to end also reveals how media strategy progresses through forgotten experiments, as failures still generate widely embraced ideas in the future.

The ephemerality of corporate-owned digital culture complicates our ability to do this historiography. On the one hand, like many digital phenomena, Social TV generated incredible amounts of archival material ripe for examination. On the other hand, this micro-moment shows how quickly Hollywood and Silicon Valley stop allocating resources to a platform or initiative, whether they orphan content on a rejected platform or erase a product altogether. Those interested in modern failure are charged with capturing the requisite material before it disappears. Although access to reliable archives is always a central challenge for historians, the archives of recent history are simultaneously vast and susceptible to variation, corruption, and erasure.[74] Even defining recent history is a fast-moving target, often with only a residue of meaningful material available. Ephemeral historiography attends to the verities of failure by situating individual utterances of failed ephemera in the proper contexts.

This book models the ephemeral historiography method by synthesizing the histories of a series of related phenomena within the micro-history of Social TV. Each phenomenon is located within its cycle of legitimation as well as linked to the trajectory of Social TV. This method is, fundamentally, a discursive analysis. Through examination of press coverage, publicity material, second-screen content, social platform interfaces, television episodes, and viewer participation, the book maps the micro-history of Social TV as it was constituted—and later discarded and transformed—by the media industries. Due to the ephemerality of Social TV, this investigation also relies on my personal activity, along with my curated archives of digital material now exiled amid corporate restrategizing or restructuring. This is not to say that one individual's navigation of a social platform or two-screen app is representative of *all* Social TV activity. Instead, this examination of Social TV consistently considers both the affordances and limitations of the archive. I utilize my experiences to enhance the press accounts and ephemeral remnants of Social TV. I find that, despite the personalized and disjointed nature of Social TV products, the discourses surrounding those products consistently propagated a default viewer orientation of collective, unified, and engaged fandom.

By highlighting the circulation of Social TV discourses on ephemeral platforms, promotional materials, and press reports, this project is part of Jonathan Gray's "off-screen studies," which asks us to contend with the material that guides how people interpret and discuss cultural products.[75] These surrounding materials are known as "paratexts," or those texts that prepare audiences to experience other texts.[76] Much of the Social TV ephemera I describe offers an exceptional vision of paratextuality as it is less "off-screen" and more multi- or simul-screen in nature, as detailed by Karin van Es and Sherryl Wilson.[77] This material was designed to fill spaces (web browsers, app windows, platform interfaces) separate from, but related to, the television screen; it also tried to structure interpretations at the *exact point* of consumption, rather than before or after release.

In accounting for Social TV's life cycle, this book exhibits one way that media studies can address recent histories of transitory culture. It is a response to recent calls for more specific examinations of new media environments and how audiences are constructed and instructed to watch television within these environments.[78] Moreover, ephemeral historiography is a call to better attend to the flash moments, temporary cottage industries, and fizzled trends that the media and tech industries hope we ignore. These ephemeral histories fall between triumph and flop, where ideas hold purchase temporarily before being shuttered when a newer idea emerges. They also, inspired by media industries studies work from John T. Caldwell and Alisa Perren and Jennifer Holt, uncover how Hollywood and Silicon Valley narrativize their relationships to innovation and failure.[79] Now, more than ever, there is too much content to analyze. But as long as corporate interests dictate the flow of information across siloed platforms, it is likely that missteps will be unceremoniously erased. Instead of only investigating the dominant players or significant trends, media and technology scholars should strive to prevent corporations from altogether diminishing these short-term experiments. Given the continued prominence of social platforms in the realm of television content creation, it is essential to analyze how Social TV first emerged, and how real-time two-screen experiences evolved into everyday social branding.

CONSTRUCTING THE SOCIAL TV PLATFORM AND FAN

The television and tech industries promoted these initiatives as innovative, interactive, or rewarding to convince people to partake in Social TV. Early trade accounts of Hollywood's interest in Twitter, for instance, framed the nascent platform as a direct line to consumers. Gavin Purcell, a producer for *Late Night with Jimmy Fallon*, attributed the show's success to its engagement with "the world's biggest test audience" on Twitter.[80] *The Vampire Diaries* producers Kevin Williamson and Julie Plec argued that social platforms were helping build "community" among not just fans but also those making television.[81] Although earnest reflections on relationships with audiences, these comments also emphasized the now-default standard of consumer behavior as an always-already connected fandom. Sharon M. Ross argues that behaviors once considered abnormal—creating and sharing amateur productions, deep speculating on plotlines, participating in fan events—have been normalized over the last twenty years.[82] As Henry Jenkins has explored, audiences are now more visibly active and equipped with creative production tools, rather than simply reactive to media products.[83] Simone Murray argues that media companies must be more responsive to these audiences to retain their loyalty in the hyper-fragmented marketplace.[84] But according to Paul Booth, the mainstreaming of fandom is due in part to media companies' perfection of "convergent incorporation," or "industry-specific discourses of immersive 'hailing' with the interactive and polysemic practices of individual viewers."[85] For Mel Stanfill, this discursive hailing reinforces norms of media use and seeks to make fans "more useful and more controllable."[86] Louisa Stein contends that the construction of the "millennial fan" as tech-savvy and mobile tries to minimize any "unruly" or "unusual" consumer tendencies and ultimately promotes fandom as "willing consumerism."[87] Fandom is the default position and good fans consume content at industry-approved times, on industry-approved platforms, and within industry-approved contexts.

The media industries circulate discourses of a very specific type of good fan: straight white men. Through a survey of popular press accounts of Comic-Con and Disney's purchase of Marvel, and Major League Soccer promotional events, Stanfill shows that men are both directly and indirectly framed as the default fan.[88] Suzanne Scott argues that the ongoing commitment to the "fanboy" archetype reaffirms Hollywood's interest in the young male demographic as the unspoken target audience.[89] Kristen Warner and Rebecca Wanzo contend that the unmarked norm of whiteness, even in assessments of female fans, requires that fans of color must carve out their own spaces.[90] While Stanfill and Scott show how this default male identity is particularly entrenched in realms like sports, comics, and gaming, television's elevation into a "quality" medium—what Michael Z. Newman and Elana Levine call "legitimation"—has coincided with similarly gendered norms.[91] For instance, HBO's "Not TV" promotional blitz and male auteur-focused discourses have positioned the premium network against the more feminized conventional television.[92] Subsequent programming and branding strategies of HBO competitors on cable have, as Taylor Nygaard and Jorie Lagerwey argue, similarly situated "the best of the medium with the masculine over the feminine and the elite over the mass, thus reinforcing cultural hierarchies that suggest women, their interests, and their stories lack importance."[93]

Social TV rarely upended television's unmarked norms and cultural hierarchies. Scott suggests that although digital platforms have amplified the voices of minority fans, these groups are still not considered essential to industry audience management strategies.[94] Sarah Florini and André Brock Jr. have recently examined how the tech industry's neoliberal infatuation with individualism generates similar "color-blind" assumptions that are baked into social platforms like Twitter and discourses about users.[95] Rather than recruit specific groups, prominent Social TV initiatives demonstrated that the identity that mattered most was that of the fan. This manufactured identity focused less on the "explicitly fannish" behaviors often associated with female fandom and prioritized more the "banal," "causal," or "phatic" activity that occurs as part of a routine of multitasked viewing.[96] But this is not a matter of taking a liberal view of fandom as an inclusionary measure. Instead, corporations exploit this reality by locating social fandom as both personalized/distracted and collective/attentive. Networks lured fans to Twitter with promises that they could participate in meaningful conversations with treasured actors or showrunners. Networks also drove fans to second-screen mobile apps with promises of behind-the-scenes access and explanatory bonus features. New market entrants, meanwhile, pushed fans to their platforms with pledges to reward them in stickers, exclusive titles, and gift cards. Each of the assurances exaggerated the utopian, participatory elements of the respective platforms and minimized references to how the corporations might benefit from fan activity.

The technological muscle of these platforms cannot be understated. As Tarleton Gillespie theorizes, platforms present as neutral spaces for users to upload and generate content for personal and collective satisfaction.[97] Yet, as we have seen over the past decade, the sophisticated power of platforms is not impartial or without hazard. The algorithms and automation that drive social platforms surpass detailed search results and personalized recommendations. They structure what we see online, and what we come to believe is important. The visible interfaces of a screen are just front-end access points for what José van Dijck calls "technical protocols," or programmed rules that govern user activity.[98] According to Lev Manovich and Alexander Galloway, platform interfaces obscure technical protocols through familiar cultural forms and visual metaphors like the representation of likes as thumbs-up or hearts.[99] The cumulative effect of these processes is that platforms and interfaces are built to be interactive, meaning they respond to user inputs in real time and over time. But technological interactivity does not enable sincere participation or influence for users. The popularity of interactive features such as a retweet or like button only reveals how platforms rely more on the endless circulation of content and less on the meaning of that content. The combination of connotative abstraction and technological control works to inspire the use of social platforms, whereby user activity is routed into financial and cultural significance for corporate power.

The legitimatizing of Social TV, then, required the convincing of several constituencies. Consumers were sold an innovative, communal, and, pointedly, *fan* multi-screen experience. Potential sponsors were promised access to consumer data that could be sorted and translated into future branded content and promotional campaigns. Tech partners were granted more considerable influence in Hollywood's promotional machinery that only further legitimized the social media ecosystem as central to modern strategy. *Social TV* explores how the resulting multi-platform content pledged behind-the-scenes access, exclusive giveaways, and real-time conversation that was subsequently controlled and circulated by corporate social media accounts.

CHAPTER SUMMARIES

I trace these ephemeral histories across different sectors of the television industry from 2010 to 2018. The case studies include a broadcast network (ABC), a basic cable network (AMC), a pay cable network (HBO), start-ups ("check-in" firms GetGlue, Miso, and Viggle), and a studio with a streaming video portal (Amazon Studios). The shadow of Netflix, and the streaming video arms race, looms large in the book, especially as Social TV products began floundering in the marketplace. The sequencing of the cases is not chronological but simultaneous. Just as they surfaced at similar times—responding to or competing with one another—the cases overlap throughout the book. The echoes underline that corporations return to the same strategies time and again, even as they promise innovation. Though each Social TV initiative emerged within specific corporate or platform conditions, my analysis reveals a persistent message of fan empowerment through the hazily defined social engagement. The commonalities across cases also speak to the narratives that the media and tech industries tell about themselves. Failure is just as easily rebranded as a pivot as it is banished deep within the digital ether.

This framing of Social TV's history is situated within an American context. While companies around the world have experimented with two-screen products, the broader discourses surrounding Social TV emanated from US entities, including the Hollywood trades. It is also worth noting that the focus on Social TV as a discursive phenomenon minimizes the focus on the voice of specific fans and fan culture within social platforms. I concede that fans utilize social platforms in ways that do not align with the strategies pursued and preferred by the media industries. These fan behaviors are important for understanding the modern social ecosystem. But I instead stress that Hollywood and Silicon Valley have appropriated fandom as the norm for all audiences, no matter how active or attentive they are.

Chapter 1 explores a prominent beneficiary of television-social media partnerships: ABC's Shonda Rhimes dramas *Grey's Anatomy*, *Scandal*, and

How to Get Away with Murder, known best by their unified branding, #TGIT (or Thank God It's Thursday). Surveying social media ephemera and more conventional publicity materials, I explore how the live-tweeting of *Scandal* inspired ABC to develop a multi-platform campaign that centered fan chatter and enthusiasm. As live-tweeting campaigns, on-screen hashtags, and celebrity participants directed audience attention across screens, ABC underscored the aura of a collective and synchronized television experience. The melding of television and social media formed a *simulated* liveness and flow that presented as immediate and communal despite being hypermediated and individualized. In charting the response to #TGIT, the chapter also reveals ABC's efforts to appropriate seemingly genuine conversational crossovers among #TGIT stars on Twitter into universe-building television spots. The construction of a promotional Shondaland universe shows how even the biggest Social TV successes were quickly incorporated into familiar Hollywood tactics.

Whereas #TGIT tried to harness the power of viewer chatter, chapter 2 examines how cable network AMC leaned more on its scripted programs to develop a branded two-screen experience called Story Sync. The Story Sync content appeared on mobile devices during live episodes of dramas like *The Walking Dead* and provided added narrative context, interactive quizzes, and games. The chapter places Story Sync into a history of content repurposing to illustrate how AMC used the app to frame its series as immersive storyworlds requiring multiple screens of content and to recruit viewers into a casual, gamified environment. While both approaches celebrated fan expertise, they couched that expertise in *synchronized reiteration*, or the consistent engagement with approved readings of themes and character actions. However, in tracing Story Sync's eventual failure, this chapter underlines that Social TV products built on something other than conversation struggled to gain purchase with viewers and industry constituencies.

The first two chapters consider how venerable television institutions navigated the Social TV realm, but chapters 3 and 4 investigate the practices of tech companies celebrated for their allegedly disruptive approaches. Chapter 3 concentrates on a group of start-ups that promised rewards to viewers who digitally "checked in" to whatever they were watching. Mirroring then-popular platform Foursquare, which enabled users to check in at physical locations and win digital badges and titles like "mayor," GetGlue, Miso, and Viggle offered stickers, monikers, and consumer goods to active participants. Situating check-ins as an extension of fan productivity, I advocate for *social productivity* that accounts for the ease at which people engage in liking, sharing, and checking in. My investigation into these now-marginalized products discloses different strategies in how each product was promoted, distributed rewards, and courted their users. While GetGlue and Miso mimicked popular social platforms and recruited fans into a collective community experience, Viggle promised a commodity-driven, individualized experience. Drawing upon my personal check-in history, observations of archived material, and an interview with a former Miso and Viggle employee, I find that these platforms aimed to remix the logic of collaborative gift economies into a corporate-friendly *reward economy*, where consumers were given small-to-moderate benefits for merely consuming media content.

Chapter 4 turns to the streaming video realm, where Social TV initiatives were not foreign despite the on-demand nature of streaming content. I detail the rise of Amazon Studios, the production company of mega-retailer Amazon. The unit promised to rattle Hollywood by uncovering amateur voices and empowering its active users to critique and workshop projects in progress. The centerpiece of Amazon Studios's development process was a public showcase for nascent television projects, where viewers could watch pilot episodes and then review them through questionnaires, star ratings, and social media posts. Utilizing publicity materials, public reporting on feedback, and my participation in Amazon's Pilot Season, I show that the company appropriated the discourses of participatory culture for a type of *fansourcing* only to gain attention in Hollywood. But despite these fan-first promises, viewer feedback had a decisively murky influence on the studio's decision-making. By tracking additional iterations of the Pilot Season approach, I contend that the studio chased acceptance from critics and the press through legitimating quality TV discourses about auteurism, creative freedom, and social responsibility.

Chapter 5 charts a critical evolution of Social TV by returning to Twitter. Although the platform was essential to the television industry's attempts to protect live television, Twitter eventually became a space for a different type of audience management. To track this shift, I examine the Twitter activity of HBO across two periods in 2014 and 2017. Drawing from a cache of tweets and research into branding, the chapter marks HBO's pivot from prestigious programmer to omnipresent content machine. I contextualize this pivot within the evolution of corporate Twitter use and HBO's growing competition with Netflix. Both developments, I argue, incentivized HBO to develop a kind of *platform authenticity* that embeds the brand more deeply into everyday life through a more diverse range of content, from playful engagement with followers to political activism. HBO's shift exemplifies that platform authenticity is part of a broader brand-creep into everyday life, where holidays and the workweek doldrums become eventized through hashtags and memes.

I conclude by considering directives by Twitter and Facebook to distribute original video content. These announcements again crystallize how television—and television industry strategies—are continuously remediated into newer platforms. They also point to a logical maturation of Social TV. Rather than serve as mediators between the industry and the audience, and between multiple screens, Facebook and Twitter have sought to unify the Social TV experience into one space and on one screen. In this regard, there is a flattening of television programming, social chatter, memes, and sponsored posts into an incessant stream of content. Social TV lives on through this deluge of ephemera, giving media brands even more prominence within the multi-platform ecosystem.

RETWEETING, FOR TELEVISION

In the spring of 2017, Ellen DeGeneres was dethroned. Her record-breaking selfie featuring some of Hollywood's most recognizable faces was surpassed by a tweet from a sixteen-year-old from Nevada. On April 5, 2017, Carter Wilkerson tweeted at the corporate account for the fast-food chain Wendy's, "Yo @Wendys how many retweets for a year of free chicken nuggets?" When the company account responded with "18 million," Wilkerson then urged Twitter to "HELP ME PLEASE. A MAN NEEDS HIS NUGGS." The random exchange inspired immense chatter on Twitter, with millions of users fulfilling Wilkerson's request and asking their peers to do the same.[100] While his tweet did not spread as quickly as DeGeneres's Oscar photo, Wilkerson became a micro-celebrity, doing interviews with news outlets and reveling in his newfound stature online.

Wilkerson's publicity tour also took him, naturally, to television. During the tweet's march toward DeGeneres's record (which Wendy's confirmed as the real mark to hit), Wilkerson appeared on *The Ellen DeGeneres Show*, the syndicated daytime talk show of his faux rival. Playing up the tension between the two, DeGeneres put her tweet on the screen and declared, "*This is a cultural phenomenon that I have right there.*"[101] DeGeneres asserted that anyone who retweeted Wilkerson's plea for nuggets must also retweet her photo and secure its lasting popularity. However, the host was eventually defeated by the enterprising teen in May 2017, aided by his appearance on her show. After crossing 3.5 million retweets, Wilkerson's victory lap included an interview on *Today*, where he playfully gloated and discussed one of the gifts he received from DeGeneres during his visit: a television.[102]

The viral popularity of Wilkerson's tweet illustrates the scope and power of social platforms. It also underlines the oddity of users talking directly to corporations that are willingly participating in the conversation to generate attention from the public. Still, the teen scored his fame in part by appearing on television and *then* returning to television to celebrate, thereby affirming the relationship between television and social media. Phenomena from one space migrate to another, and back again, creating a circulation of ephemeral content, fame, and chatter. This book investigates these relationships, flows, and circulations to show how television and social platforms rely upon and remediate one another.

Chapter 1

FROM TGIF TO #TGIT

Simulated Liveness and Flow in Shondaland

To celebrate the start of the fall television season in September 2015, Twitter introduced exclusive series-specific emojis for three Shonda Rhimes-produced ABC dramas: *Grey's Anatomy*, *Scandal*, and *How to Get Away with Murder*. While Twitter always allowed users to click on hashtags—e.g., #Scandal—to follow ongoing conversational threads, the emojis added a visual wrinkle to the experience. Tweets sent with #Scandal generated cartoon versions of characters Olivia Pope (Kerry Washington) and Fitzgerald Grant (Tony Goldwyn) blowing kisses at one another as a heart floated between them. The use of #GreysAnatomy spawned a red heart covered with a bandage, while posts tagged #HTGAWM produced the Lady Justice statue central to the series' early storylines.[1] Known as "hashflags" due to their emergence during global events like soccer's FIFA World Cup, Twitter's custom emojis exist only for select periods to accentuate the liveness and immediacy of associated events. This temporally restrictive promotion costs a premium; participating companies paid $1 million for their custom promotional emoji during the Super Bowl in 2016—five times the cost for a conventional "Promoted Trend" that inserts a sponsored message into the user-generated list of trending topics.[2]

In detailing ABC's partnership with Twitter, executive Ben Blatt offered a fan-centered rationale: "Anytime something emotional happens, (fans) are using emojis to express that emotion. We wanted to make sure we kept rewarding them with fun new ways to inspire them to do that."[3] With the emojis activated in a window around live Thursday broadcasts, *Grey's*, *Scandal*, and *Murder* topped Nielsen's Twitter TV Rating chart for the first week of the fall season. Approximately 10 million Twitter users sent more than 1 million unique tweets across the three hours.[4] The creation of ABC's Twitter emojis was yet another example of the enormous social footprint of Rhimes's three series. Though *Grey's* had been one of television's most popular offerings since its 2005 debut, *Scandal*'s rise from modest Nielsen performer in season one to cultural phenomenon by the end of season two happened in part due to the cast and crew's willingness to live-tweet with viewers, turning each new episode into a must-see and must-tweet event. *Scandal*'s emergence—first on social media and then among the broader television audience—was so significant that ABC made the historic decision to give an entire night of the primetime schedule to Rhimes to coincide with the *Murder*'s debut in the fall of 2014. To promote the trio of dramas on Thursdays, the most lucrative night for advertising dollars, ABC unveiled a synergistic and Twitter-centric campaign identifiable by a hashtag: #TGIT, or Thank God It's Thursday.[5] To a degree, the tagline was a self-referential nod to ABC's TGIF (Thank God It's Friday) branding of the 1990s and apiece with other multi-platform promotional campaigns within the ABC family of networks.[6] But more pointedly, the #TGIT campaign is notable for how it was built with both old (print ads and posters, television spots, contests) and new (tweets, viral-baiting hashtag mysteries, memes) promotional tactics, all of which stressed the emoji-worthy twists and tweetable moments of Rhimes's series.

At the time, the success of #TGIT contradicted some of the talking points about the potential demise of broadcast television. The broadcast networks survived similar pessimistic projections during the advent of home video and cable television, albeit with a smaller share of overall viewership. But between the higher saturation of DVRs, the increased number of digital video marketplaces and streaming video portals, and the uptick in original series production for those new distribution channels, broadcast television—especially *live* broadcast television—faced an even greater fragmentation of its previously "mass" appeal. This reality motivated industry leaders and sponsors to push for new data that could sustain revenue for broadcast television, including those that tabulated viewership as far out as twenty-eight days after live airings.[7] There was a growing belief that modern viewers did not care about preprogrammed schedules or appointment television in the live primetime environment.

Scandal hashflag featuring characters Olivia Pope and Fitzgerald Grant.

But #TGIT showed that broadcasters could still generate attention for primetime programming in the Social TV era. This chapter investigates how Rhimes's programs and the #TGIT campaign became such television and social media marvels. Drawing upon examples from Twitter and industry discourse, I explain that *Scandal* thrived as a Social TV artifact due to the enthusiastic participation of Rhimes and star Kerry Washington but also the broader troupe of actors, writers, and producers from Rhimes's series. Networks and producers had irregularly experimented with live-tweeting new episodes (including for other Rhimes projects), but the *Scandal* cast provided a consistent and conversational front that sustained interest, unlike sporadic sponsored "Tweet Weeks." Naturally, ABC tried to convert the social media fervor surrounding Rhimes's programming into more conventional publicity materials, even as the #TGIT campaign centered on hashtags, live-tweeting, and fan enthusiasm.

As the campaign tried to convince viewers that they must tune in live on Thursday nights or risk missing the big twists, juicy hookups, and star-studded communal experience, I argue that it also tried to reinforce the immediacy of broadcasting, where liveness and flow directed audience attention across act breaks, episodes, and, most crucially, screens. My analysis reveals how this reconstituted liveness and flow—key ontological features of television that have long been used to structure the viewing experience—manifested on television via chyrons and interstitial teasers and on social platforms via hashtags, exclusive emojis, photos, videos, and celebrity chatter. As ABC aimed to direct the multi-screen experience, the network fortified the relevance of social chatter by championing it all across television. ABC reminded everyone that stars and fans were tweeting along *right now*, cited the high volume of social impressions in commercials, and crafted self-reflexive jokes about the growing dominance of Rhimes and her linked universe of soapy serials.

#TGIT's melding of television and social media produced what I call *simulated* liveness and flow—a coordinated, quasi-immediate experience mediated by multiple screens and multiple platforms that was nonetheless presented as uniquely cooperative and conversational. This simulated liveness and flow was designed to feel like an idealized version of "co-connected viewing": simultaneous, smooth, and fluid.[8] Viewers were guided from episodes on the television screen to the social chatter on the second screen, where the linkages between material constructed an immediacy and liveness. The chyrons and interstitials that once sold television's liveness now worked to funnel viewers to what Ruth Deller calls preprogrammed "markers of communality," targeted corresponding hashtags, real-time tickers displaying popular posts, and reminders about engagement with celebrities.[9] This multi-platform strategy aimed to *eventize* #TGIT, or surround it with an aura of ritual, community, and attention more historically associated with Daniel Dayan and Elihu Katz's live "media events" like awards shows or global sports telecasts.[10] While, as Jérôme Bourdon argues, media events offer "maximum liveness" and feelings of togetherness, the success of #TGIT reveals that the desire to be part of a sweeping, real-time conversation and community extends to scripted television as well.[11]

Yet, the focus on #TGIT as a live multi-screen event, with the active participation of many #TGIT cast members and fans, also ignored the realities of the personalized, distracted, and inelegant navigation between screens and instead promoted an imaginary collective, attentive, and cohesive happening for all viewers. Indeed, as José van Dijck writes, "the tweet flow," is constituted as "a live stream of uninhibited, unedited, instant, short, and short-lived reactions."[12] In this case, ABC packaged the uninhibited tweet flow as part of the overall #TGIT experience, accelerating Nick Couldry's "myth of a 'shared' ritual center."[13] Despite the acutely personalized process of watching, live-tweeting, and navigating between screens, viewers were still pushed by industry discourses to *feel* more connected than ever.

Amid the discourse about accelerated time-shifting and the end of appointment viewing, the success of *Scandal* and #TGIT proves that broadcasting could be amplified by social media rather than destroyed by it. Having first established the import of Rhimes's auteur persona, this chapter then examines liveness and flow as tools for audience management. Pre-*Scandal* strategies from ABC and Rhimes indicate that the #TGIT campaign represented a cunning progression of liveness and flow, in conjunction with social media, to appeal to Rhimes's fans. Next, I examine the *Scandal* cast's embrace of Twitter as a space for fan outreach and ABC's ensuing opportunistic co-opting of this outreach within the coordinated multi-platform publicity initiatives that also utilized hashtags. In charting Rhimes's persona and fan engagement through the creation of #TGIT, this chapter illustrates how Twitter chatter boosted the conventions of her work and her promotion of apolitical multiculturalism. Yet, as Rhimes, one of Hollywood's most prolific Black creators, avoided specific conversations about representation, ABC used the diversity of #TGIT to court Black viewers and "Black Twitter," the influential sub-section of the platform made up of Black users, without openly challenging racial normativity. Similarly, by identifying the intermingling of #TGIT stars—first on Twitter and then print ads and television spots—I exhibit how social media novelties are subsumed into the standard Hollywood promotional machinery. These conversational crossovers helped build anticipation for Thursday night social talk but also constructed a universe full of Rhimes-affiliated stars personified by the name of her production company, Shondaland.

@SHONDARHIMES: SHONDA RHIMES AS SOCIAL AUTEUR

While the success of *Grey's* catapulted Rhimes into superstardom, she maintained an active relationship with her fans online, initially on an ABC blog and then on Twitter. When *Variety* first reported on showrunners using social media in 2010, Rhimes was identified as a "Twitter maven."[14] Her affable persona on Twitter manifested in familiar complaints about the workweek and her endless battle with productivity:

> @shondarhimes: Oh, it is Wednesday already. How is the week going so quickly? I have too much writing to do. Time needs to slow down so I can catch up. (January 6, 2010)

A week later, Rhimes fielded a question from a fan about not being "verified," Twitter's marker for popular users. Rhimes responded with humor:

> @shondarhimes: Twitter decides if u are super cool enuf to need verifying. I clearly am not super cool. (January 13, 2010)

The tweets positioned Rhimes as a typical user unsure of the value of social media but still happy to use it as a distraction from other responsibilities. By addressing her lack of verification, Rhimes expressed the desire shared by millions of people looking for recognition on Twitter. She also delivered a self-deprecating comment on her status as a relatable showrunner, the catch-all term used to describe the managerial and creative functions of a writer/producer.

While showrunner has a practical meaning, it also, as Michael Z. Newman and Elana Levine argue, operates as a discursive tool in the promotion and reception of television.[15] As new outlets for promotional material and communication coalesced with the expansion of television criticism in the 2000s, some showrunners became celebrity figures.[16] While digital media gives fans more access to celebrity showrunners, Alan Wexelblat claims that producers have traditionally used emerging tools to formalize official story explanations or to delegitimize rogue fan perspectives.[17] For instance, Ron

Moore (*Battlestar Galactica*) and Damon Lindelof and Carlton Cuse (*Lost*) relied on DVD and podcast commentaries to justify production decisions and to answer fan questions in ways that influenced future interpretations of their series.[18]

Though her position is now "undisputed," Rhimes is not the typical celebrity showrunner.[19] Lauded for the diversity of her productions, she became a proponent of color-blind casting, a purportedly democratic process that creates roles to be filled by the "best actors" no matter their race.[20] Rhimes's reputation inspired many stories about how her identity shaped her work and position as a boss. Citing a blog post from writer Krista Vernoff, a 2006 *New York Times* profile recalled that prior to *Grey's*, Rhimes had not been in a writers' room and nervously stood outside the door before the first day of production.[21] But instead of framing Rhimes as a solitary visionary, the profile charted a new eagerness to collaborate. Producing partner Betsy Beers confirmed that while Rhimes's confidence in her voice had grown, it was her new ability to lead others that showed necessary personal and professional maturation. Anecdotes like this stressed Rhimes as the creative force of *Grey's*. Still, whereas male producers are often elevated via geek credentials, literary influences, or autonomous genius, Rhimes was framed as an inspirational leader.[22] The mention of a blog post from another writer nodded to Rhimes's open sharing of her life with fans and her egalitarian vision of writers' room communication. Together, these anecdotes underscored the gendered nature of auteur discourse.[23]

Coverage commonly attributed the diversity in *Grey's* and its matter-of-fact handling of race to Rhimes's identity. Another *New York Times* story stressed that *Grey's* had "differentiated itself by creating a diverse world of doctors—almost half the cast are men and women of color—and then never acknowledging it."[24] Then-ABC president Stephen McPherson linked *Grey's* representation to its popularity among the "diverse canvas" of American viewers.[25] Rhimes, meanwhile, negotiated the racialized contexts in which agents of auteurist discourses placed her. She said that while *Grey's* was more representative, race did not drive her writing of storylines. "I'm in my early 30s, and my friends and I don't sit around and discuss race. We're post-civil rights, post-feminist babies, and we take it for granted we live in a diverse world," she said.[26] Pressed to explain her view of race as it pertained to casting, Rhimes declared that she wanted *Grey's* "to look like the world," but she also affirmed that she hired the "most talented" actors, separate from race.[27] Newman and Levine argue that "It helps us understand television authors' creative functions when we learn that their own lives become fodder for storytelling."[28] In this case, coverage wanted Rhimes to stress her life experience in the racial context to ensure that her series came from an authentically racialized place. Rhimes pushed back against these constructed assumptions and instead promoted a post-everything style of pluralism.

Rhimes's attempts to counter the framing placed upon her by the press exemplified the challenges historically facing minority creatives in Hollywood. US television has, at best, a mediocre track record of wrestling with sociocultural issues of the moment. Herman Gray argues that the industry's systems of production reify racial stereotypes and hierarchies that produce flawed nonwhite representations.[29] Gray claims that Black representation takes on what he calls an "assimilationist" perspective by making the historical and contemporary impacts of racial inequality invisible or inconsequential to the characters.[30] Meanwhile, minority viewers are often only pursued or valued by networks looking for a mark of differentiation. For instance, Fox's arrival in the late 1980s briefly improved minority representation. Kristal Brent Zook argues that Black-produced series like *In Living Color* and *Living Single* presented as authentic through their use of autobiography, improvisation, and pride in Black visual signifiers.[31] But once Fox attracted more white viewers and sponsors with *The Simpsons* and procured the rights to professional sports, the network phased out the Black-focused hits. In the late 1990s, a pair of nascent networks, The WB and UPN, also filled their schedules with series from Black artists. Though they were less successful than Fox, the tactic temporarily buoyed the networks before they predictably pivoted away from Black audiences. The cases show that, as Beretta E. Smith-Shomade argues, Black programming has been progressively "migrated, or otherwise exiled" to the fringes of television *unless* it is needed to court Black audiences as part of a temporary niche branding strategy.[32]

On the one hand, Rhimes avoided speaking for all minorities or trying to change Hollywood's structural racism. Rhimes's status as one of the few Black showrunners set her up for increased racialized scrutiny. To combat this, Ralina Joseph contends that Rhimes treated race and racism with "strategic ambiguity": "*strategic* in that it is a mindful choice; it is *ambiguous* in that it deploys a primary facet of post-race, *not naming* racism."[33] By claiming to use color-blind casting that *just happened* to produce a diverse group of actors, Rhimes endorsed multiculturalism without framing it as the outcome of a political agenda or a challenge to societal norms. On the other hand, Rhimes's post-racial approach did not engage with racial inequality. While Joseph submits that Rhimes's ambiguous discourse enabled her to traverse institutional barriers, Kristen Warner argues that the effect of Rhimes's approach is a surface vision of difference that is comforting to white viewers at the expense of specific representation of people of color. For Warner, Rhimes is smart to avoid television's tendency to situate race into the box of a "very special episode." However, she still promotes an idealized belief that "representation alone can change preconceived notions about racialized groups."[34]

Indeed, it is Rhimes's approach that helped her become a symbol of Hollywood's self-satisfying discourses about progress without disrupting the existing structures that impact many minority creatives. Rhimes advanced this approach with her Twitter performance, where she revealed her anxieties, expressed her fandom for other series, and shared in fan appreciation for the talent and appearance of her programs' stars. Lisa Schmidt argues that "fanboy auteurs" like *Supernatural*'s Eric Kripke embody a "personal and chatty" voice that addresses the fan as a friend even though the relationship is illusory."[35] Social platforms, where Rhimes made herself available and relatable, increase this perceived intimacy. She even declared on *Good Morning America* that she enjoyed reading fan tweets in contrast to the solitary alienation of writing.[36] This performance of auteurism helped Rhimes navigate the industry and the vitriol of social media conversation. But by framing her Twitter experience positively, Rhimes also encouraged fans to use the platform to talk about her projects and set the stage for future social media initiatives.

More pointedly, Rhimes's approach enabled ABC to position her projects as what Maryann Erigha calls "crossover phenomena" that appeal to multiracial audiences without a racialized specificity that requires those audiences to imagine themselves as part of a particular group.[37] While Rhimes continued to assert that her work was more about the interpersonal predicaments of her characters, ABC could take credit for empowering a Black female showrunner as well as facilitating racially diverse productions and growing social chatter. This ultimately enabled ABC to leverage Rhimes's vibrant persona for significant multi-platform buzz and live viewership within an entire night of branded programming.

FROM "LIVE" TO #LIVE TELEVISION

Broadcast television has been linked conceptually to liveness and flow for decades. Liveness describes how, as Jane Feuer asserts, "Events *can* be transmitted as they occur; television (and videotape) look more 'real' to us than does film."[38] John Ellis contends that liveness produces an "effect of immediacy," as though an image on-screen is being produced, transmitted, and received simultaneously.[39] For Philip Auslander and Nick Couldry, liveness's effect of immediacy constructs a sense of intimacy with both the image and with people watching elsewhere.[40] This sensation of feeling television's liveness serves as a mark of distinction for the medium, according to Elana Levine.[41] Meanwhile, flow is famously described by Raymond Williams as television's organization into a "sequence" of linked programs, commercials, and interstitials. Williams devised the idea of flow to clarify the experience of taking in television for hours at a time, where it just appears on-screen like a never-ending flow of content unbothered by the involvement of the audience.[42]

While these elements are vital to the phenomenological experience of watching television, the industry has been persistent in its efforts to control and deploy the power of liveness and flow. The attempts have emerged at all levels of production, programming, and promotion in hopes of driving audiences to advertisers. News telecasts, sports contests, and live events construct liveness via "LIVE" and time chyrons in the corner of the screen, as well as through anchors, hosts, and talent looking directly into the camera or speaking directly to the audience—all of which stress that the broadcast is occurring at that moment. Visual continuity—a sequence of shots from a playing field to the crowd reaction to a replay and back to the field, bolstered by a real-time audio track—also helps sell the sensation of immediacy and liveness.[43] As Auslander argues, despite the move to a few-second tape delay, networks have continued to exploit the aura of liveness by selling the idea that *anything can happen* when television goes live.[44] Over time, networks have had less control over audiences, but, per Ien Ang, they have also steadily pursued "risk-reducing techniques" to enhance liveness.[45] In the 1950s, networks promoted the live nature of anthology dramas as a way to gain a foothold against local stations.[46] More recently, they have sold live episodes of typically pre-taped series like *ER* or *30 Rock* as distinctive special events.[47] Even with VCRs, DVRs, and streaming portals, networks continue to assert the power of what Feuer calls the "ideology of liveness."[48]

Similarly, flow has been consistently "planned" by those Williams calls "providers" (networks, studios, producers, and sponsors) to keep viewers from tuning out.[49] On a macro level, planned flow is the construction of series into particular timeslots, days, and seasons. This framework results in a meticulously managed schedule, where each daypart is tailored to a specific audience with relevant genres of programming (e.g., midday soaps targeted at stay-at-home mothers). On a micro level, planned flow manifests in the act-break structure of episodes. Opening sequences try to hold the audience's attention through the first commercial break, effectively serving as a trailer for the rest of the episode. Concluding cliffhangers, meanwhile, try to convince viewers to return the following week. Together, these "predetermined" elements facilitate an enthralling experience intended to attract the largest and most lucrative audience.[50] Ellis claims that the flow found in the programmed schedule is made more effective by the use of themed program blocks with catchy slogans and lead-ins, where the planned flow enables a more popular series led directly into a newer or less popular one.[51]

To a degree, each wave of new technology has curtailed the power of liveness and flow, freeing viewers from watching within preprogrammed systems. As William Uricchio argues, innovations as early as the remote control signaled a shift away from "program-based notion of flow" and toward a more "viewer-centered notion."[52] For Uricchio, this process is shaped by the applied metadata protocols, filters, and search algorithms found within set-top boxes, streaming portals, and digital stores.[53] But while technology performs this kind of flow and allows viewers to craft an on-demand personalized experience, networks and studios remain committed to evolving their approaches to retain a degree of control.

The evolution of networks' support for liveness and flow is evident in pre-#TGIT strategies for Rhimes-produced series. Before *Scandal*, Rhimes brought two other series to ABC's schedule: 2007's *Grey's* spin-off *Private Practice* and 2011's medical soap *Off the Map*. Both series were promoted on Rhimes's storytelling mettle and, at various points, linked to *Grey's* on the schedule. When *Grey's* became the lead-in for *Practice* in 2009, ABC underlined the connection between the programs with a planned crossover that also established that they would subsequently air together. Rhimes previewed the crossover by stressing the value of flow: "There's something lovely about having that much real estate on Thursday night. I feel like *Private* is getting a real chance to have a strong lead into it."[54] Trailers and anticipatory interviews like this support the flow constructed by programming blocks and crossover events.

Two years later, ABC tried to use Rhimes to sell *Off the Map*. Though Rhimes was not directly involved in the production, promotional materials for *Map* branded the series as "from the producers of *Grey's Anatomy*" to emphasize her presence. For her part, Rhimes used her Twitter platform to pitch *Map* and its creator (a former *Grey's* scribe) Jenna Bans to followers:

> @shondarhimes: OTM = OFF THE MAP. My new show. Wed night. 10 pm. ABC. January 12. Created by JENNA BANS. AWESOME WRITER! (November 3, 2010)

@shondarhimes: While I'm busy writing, check out what Off The Map Creator @jennabans is up to as she prepares for her show's premiere on January 12! (November 29, 2010)

Once *Map* began in January 2011, Rhimes accelerated her enthusiastic push that hinted to the multi-platform future of promotion for *Scandal* and #TGIT:

@shondarhimes: Grey's and Private are new tonight! Off The Map premieres next week!!! Am using too many exclamation points!! (January 6, 2011)

@shondarhimes: East Coast: Off The Map in less than five minutes!! (January 12, 2011)

@shondarhimes: Tonight is OFF THE MAP night! Tomorrow is GREY'S AND PRIVATE DAY! (February 16, 2011)

Rhimes's stamp of approval for the new series transmitted an excitement and immediacy to followers, punctuated by references to specific temporalities ("next week," "tonight," "less than five minutes"). The passion for *Map*—and her performative note about using too many exclamation points—situated Rhimes as an enthusiastic fan despite her privileged role as a producer. But the animated tweets also tethered *Map* to her more popular series, essentially performing the same associative purpose as the usual "from the producers of *Grey's Anatomy*" commercials. In the hands of an influential figure like Rhimes, this served a new kind of just-in-time promotion that pushed fans to hop from one screen to another.

Once it was clear that *Map* had underperformed with audiences, ABC tried to lure *Grey's* fans to the struggling new series with "sneak peeks" of *Grey's* episodes to come the next night. Rhimes again turned to Twitter to fuel attention in the exclusive snippets and the struggling *Map*.

@shondarhimes: Special sneak peak [sic] of GA musical during tonight's new Off The Map!!! (March 2, 2011)

@shondarhimes: I am pretty sure there is. Will check. "RT @Stephanie_Mere: Is there a sneekpeek [sic] of grey's tonight during off the map??" (March 16, 2011)

The tweets signaled that Twitter was becoming more central to the network's promotion but also show how ABC tried to recoup some of the power of liveness and the planned flow. Given the time restrictions of a sneak peek for an episode that would air twenty-three hours later, Rhimes's personalized pitch tried to create a genuine and immediate reason to watch *Map* live.

Despite Rhimes's best efforts, ABC canceled *Map* after just one season. At the time, networks were still searching for the best way to use social platforms, and synchronous live chatter was still relatively uncoordinated and experimental. Rhimes's tweets were more of an extension of her online persona than part of a coordinated campaign. But the normalization of on-screen chyrons and live-tweeting helped ABC promote simulated liveness and flow.

#GLADIATORS UNITE: ABC CO-OPTS THE LIVE-TWEETING OF *SCANDAL*

A May 2012 Associated Press report on the rise of live-tweeting began with an anecdote about actors joining the conversation. The story used Kerry Washington, star of the just-premiered *Scandal*, as its chief example. Without the involvement of ABC, Washington decided to answer fan questions during *Scandal*'s pilot episode while she watched with family. As she recalled later, "I hate watching myself. So while the show was on, I was buried in my laptop tweeting. It was fun."[55] Washington enjoyed the experience so much that she tweeted during the West Coast airing as well. That evening, Washington's Twitter feed revealed the origins of what would become a social media and television phenomenon. Though Washington's publicity team had previously controlled her account to promote *Scandal*, the star supposedly took over once the episode began. She encouraged viewers to send questions to the #AskScandal hashtag and responded to countless inquiries about her feelings on the script, production anecdotes, and plot teasers.

Washington responded to many questions by directly replying to fans, preceding her answers with a friendly "hey" and finishing with playful emoticons, exclamation points, all-caps enthusiasm, and endearingly awkward formatting. In a foundational moment, she retweeted one user's idea for what to call fans:

> @kerrywashington: "@PSawyerSchue: @ColumbusShort1 @Scandal-ABC @kerrywashington Here On Out, Us Fans Of The Show Will Be Known As #Gladiators? ;)" I LOVE THAT! (April 5, 2012)

The use of "Gladiators" came from the pilot episode, in which the political fixers led by Washington's Olivia Pope are referred to as "gladiators in suits." When a fan proposed the nickname for the fans to Washington and co-star Columbus Short, she responded enthusiastically. Just thirty minutes later, the use of #Gladiators had already taken hold, as Washington and co-star Darby Stanchfield expressed their gratitude with the term of endearment:

@kerrywashington: How's it going #Gladiators?!?! SO EXCITED THAT YOU'RE WATCHING WEST COAST! (April 5, 2012)

@kerrywashington: AGREED! "@darbystnchfld: #Scandal cast LOVE OUR #Gladiators!!! THANKS FOR WATCHING #Gladiators!!! AMAZING FANS ALREADY! #AskScandal XOXOXO" (April 5, 2012)

Within an hour, active tweeting by Washington and her co-stars ratified both the practice of live-tweeting *Scandal* and the Gladiator nickname. As with Rhimes, Washington's use of capitalization and punctuation and willingness to respond to fan queries exhibited a zeal for chatter about her series and fellow tweeters.

Scandal debuted in April 2012, surrounded by familiar industry discourses and similar strategies by ABC. Press coverage stressed the presence of typical Rhimes conventions like "sexy plotlines and the attractive multicultural cast," as well as the project's "tremendous responsibility" of representation given that Washington was the first Black female lead of a primetime series in nearly thirty years. But Rhimes asserted a post-racial perspective on her history-making lead character, who was based on Judy Smith, a former press aide for George H. W. Bush and crisis manager: "A good story is a good story."[56] Meanwhile, ABC used the still-popular *Grey's* as *Scandal*'s lead-in and promoted the auteur-focused "from Shonda Rhimes" tagline in trailers. Though most of the initial discourse about *Scandal* placed the series within the context of Rhimes's previous efforts and auteur persona, Washington's live-tweeting of the pilot gave the series a Social TV mark of distinction. But Washington also catalyzed a sense of community among the cast, crew, and fans that would be co-opted by ABC's promotional machinery.

Before *Scandal*, Rhimes promoted the live-tweeting of her series, but the practice usually only happened during significant events like crossovers and ratings sweeps periods. But as Rhimes credited in *Variety*, Washington's activity morphed live-tweeting into "a different tool" for producers and ABC.[57] During the April 12, 2012, broadcast of *Scandal*'s second episode, Washington's live-tweeting was publicly acknowledged and approved by her boss: "@shondarhimes: @kerrywashington Hi Kerry!! Am loving your live-tweeting!" This friendly message between showrunner and star suggested that there was no top-down decree from ABC driving Washington's activity. Rhimes quickly incorporated the campaign into her tweets and pointed fans toward the now-official hashtag and a custom URL for ABC's website: "@shondarhimes: Scandal cast will tweet live during tonight's epi at 10PM ET/PT. To submit questions, tweet using hashtag #AskScandal" (April 19, 2012). Washington, meanwhile, continued to candidly engage with fans before, during, and after each new episode. Washington pushed tweets to the #AskScandal hashtag but couched that directive as gratitude for and commitment to dialogue with her fans:

> @kerrywashington: Done with work!!!! Racing home to turn on my DVR and live tweet with you #Gladiators :) use HASHTAG ---> #AskScandal (May 3, 2012)

> @kerrywashington: Im SO grateful that I got off work in time to LIVE TWEET w/ @ScandalABC #Gladiators - YOU ALL ARE THE BEST! Truly! Thanks for hanging w/ us. (May 10, 2012)

Washington stressed how much she valued fan comments and the sense of togetherness produced by the ongoing live conversation. Her language underlined the positive qualities associated with liveness: immediacy, presence, community, and spontaneity. References to tweeting *with* fans or hanging *with* the cast promoted the live-tweeting experience as community-building—to the point that she had to "race home" to participate.

Washington's passion for live-tweeting inspired the participation of co-stars like Tony Goldwyn, who played President Fitzgerald "Fitz" Grant, Olivia's love interest. Many of Goldwyn's responses to #AskScandal were about his character's romantic entanglements:

> @tonygoldwyn: "@Traci_Reid: @tonygoldwyn #askscandal Is the Prez still in love with Olivia?" OH YEAH!!! (April 26, 2012)

> @tonygoldwyn: "@Missknowmyworth: #ASKSCANDAL WILL THERE BE ANY ALONE TIME FOR OLIVIA & THE PRES @kerrywashington @shondarhimes @tonygoldwyn" KEEP WATCHIN (May 3, 2012)

Goldwyn embodied the same passion displayed by Rhimes and Washington; he used both full capitalization and exclamation points to denote his excitement in responding to questions. The tweets also assured fans that *Scandal* would continue to explore the romantic tension. Goldwyn's confidence in his character's feelings was visible within early episodes that tried to motivate fans to get more invested and continue to tune in.

While *Scandal* garnered modest Nielsen ratings in its first season, the coordinated #AskScandal campaign helped each episode produce thousands of tweets. But by the first season finale, ABC began to push more obvious promotional tactics into the Twitter community. Along with #AskScandal, the network promoted a new hashtag, #WhoIsQuinn, a reference to a finale plot twist involving the secret identity of Quinn Perkins (Katie Lowes). Washington, Lowes, and the official *Scandal* account circulated the hashtag immediately during the episode and throughout the summer and fall in preparation for season two:

> @kerrywashington: Looooooooove hearing that so many of you are screaming #WhoIsQuinn at your TVs! LOL @KatieQLowes @ScandalABC (May 19, 2012)

> @ScandalABC: What are your Quinn theories? #WhoIsQuinn #Scandal (September 12, 2012)

> @KatieQLowes: LOVE reading all ur #WhoIsQuinn theories! Keep trying ;) U WILL find out 2night! Tell ur friends u won't want 2 miss it. #Scandal 10/9c ABC (September 27, 2012)

#WhoIsQuinn epitomized a shift for *Scandal* as a multi-screen social series. The new tagline tried to appropriate one of Rhimes's patented cliffhangers not just for a mystery that would keep the audience talking throughout the summer but also for a hashtag that would keep them talking on a specific platform like Twitter. Rather than focus on the spontaneous and enthusiastic chatter between stars and fans, the campaign aimed to channel fan energy toward a more coordinated conversation. The network's attempted co-opting of the enthusiasm surrounding *Scandal* produced a sharp contrast among the above #WhoIsQuinn tweets. The network account offered a generic question as a performance of engagement, whereas Washington and Lowe retained their passionate tone and style visible in other tweets. These vernacular distinctions expressed a tension in *Scandal*'s move from an organic Social TV sensation to a more manufactured seeker of viral attention. Indeed, #WhoIsQuinn was more than a hashtag; ABC created an ARG-like website—whoisquinnperkins.com—featuring "depositions" of Quinn's co-workers and her mugshot seen on the series. The site was framed as a rogue exposé against Olivia's firm: "Olivia Pope has her hands full with a scandal of her own. It seems the renowned fixer has something to hide." Each video teased upcoming plot twists and reiterated the season two premiere date.[58]

On the one hand, #WhoIsQuinn revealed how the combination of television storytelling, on-screen hashtags, and fervent tweeting could drive interest in a live episode and later sustain it into the next season's premiere. Rhimes trained viewers to expect major cliffhangers, motivating them to watch live and not be spoiled. Chatter between seasons only cultivated anticipation more, making the need to watch the premiere live *even more* pressing than before. Simulations of liveness and immediacy—of knowing the information and sharing the experience—were useful tools in the battle against time-shifting. On the other hand, the execution of #WhoIsQuinn embodied a familiar model of audience management and content distribution online. Those who followed #WhoIsQuinn to the website were granted access to small bits of new information but only in the form of brief teaser videos common to network publicity since the mid-2000s.[59] Engagement with fellow fans was funneled through faux-interactive content. The corporate strategy did not effectively convince more people to watch live. The season two premiere of *Scandal* had lower ratings than most season one episodes.[60]

Despite this potential misstep, the social chatter around *Scandal* continued to grow in the fall of 2012. During a pivotal moment of the November 29 episode—the attempted assassination of Goldwyn's Fitz by an unknown culprit—ABC debuted another mystery-oriented social campaign. The response to the cliffhanger was predictably intense, as Washington nurtured the shock by retweeting nervous fans and performing her all-caps excitement:

@kerrywashington: "@JuliaOnTV: @ScandalABC is going to send me into cardiac arrest soon. Heart is beating mile a minute. #askscandal @shondarhimes" BREATHE :)

@kerrywashington: #askscandal YES! #GLADIATORS are going CRAZY! We're loving it! FELT THE SAME WAY WHEN WE READ IT! CRAZY! Thanks for watching! We love youz!

ABC, however, rapidly directed fans to a new hashtag, #WhoShotFitz, a reference to another famous network promotional campaign, "Who Shot J. R.?" from the CBS soap *Dallas*. Both Washington and the @ScandalABC account immediately started using the hashtag. The latter also circulated an image of Fitz in the design of Shepard Fairey's famous poster featuring Barack

Obama with "#WhoShotFitz" awkwardly in the place of "HOPE." But the network seemingly learned its lesson from #WhoIsQuinn. Rather than try to push fans to separate websites or ancillary video content, #WhoShotFitz was entirely centered on social media chatter. The official press release declared that the campaign would help fans "explore the theories and questions behind the storyline" on Twitter as the cast live-tweeted December episodes.⁶¹

#WhoShotFitz demonstrated the new potential of a social media-driven multi-platform campaign. Conventional trailers for both online and broadcast audiences teased that "the gunman will be revealed." The hashtag appeared on-screen during subsequent episodes as the cast solicited fans' most complex theories during the live-tweeting conversation. And ABC continued to circulate illustrations of Fitz and the hashtag. This material worked in concert for maximum social shareability and to build anticipation for new evidence and new episodes. While the focus was on spontaneous social media chatter, ABC still used this material to guide audience attention from screen to screen and generate that chatter—both in the live broadcast environment and between each *Scandal* episode.

The synergy of promotional content and platforms constructed simulated liveness and flow. ABC pushed viewers to tune in live to equally revel

ABC promotional image for the #WhoShotFitz campaign mimicking Shepard Fairey's famous "Hope" artwork celebrating Barack Obama.

in the resolution of the cliffhanger and the reactions from fellow fans and *Scandal* stars. Viewers were directed from the preprogrammed television flow to the algorithmically programmed Twitter. Although, as José van Dijck claims, Twitter's flow appears more "uninhibited, unedited, instant, short, and short-lived," the platform still structures what users see and how they experience the real-time timeline.[62] In a coordinated campaign like #WhoShotFitz, tweets from verified or popular accounts like @ScandalABC, Washington, or Rhimes will be most visible. Hashtag initiatives like this one also construct an imaginary universal collective experience that promotes the false idea that everyone is watching and tweeting along in the same way at the same time—one approved multi-screen flow experience for all. In trying to structure the online conversation around specific pre-approved talking points, #WhoShotFitz capitalized on the genuine community emerging around live episodes of *Scandal*. While this strategy did not necessarily minimize the value of the shared collective experience of watching and live-tweeting, it exhibited how networks were beginning to understand how to incentivize people to participate in that experience in ways that most benefitted corporate objectives.

#WhoShotFitz also proved that synergistic hashtag campaigns and simulated liveness and flow could be successful. *Scandal* scored record ratings in the 18–49 age demographic and total viewers for episodes promoted around the attempted assassination of Fitz. The December 13, 2012, installment, which revealed the shooter's identity, also drew 2,838 tweets per minute and nearly 158,000 across the hour.[63] When *Scandal* returned from its holiday hiatus in early 2013, ABC more consistently promoted #AskScandal on its airwaves. Meanwhile, *Scandal*'s live ratings and social footprint continued to grow, as did press coverage about the cast and fans "watching together."[64] The *Hollywood Reporter* pointed to Twitter as key to *Scandal*'s "sophomore growth spurt," with a recent episode producing five worldwide trending topics and the highest-ever retention of the *Grey's* audience.[65] An *Entertainment Weekly* cover story stressed *Scandal*'s appeal as a "live-tweet-every-*oh-my-God-moment* viewing experience."[66] *Slate* produced weekly "tweet-watch" posts, collating tweets from actors, fans, and critics.[67] Twitter, always angling to collaborate, partnered with Stanchfield to collect "the best *Scandal* Tweets of the night" in one place, and cross-promoted the event on its official blog.[68]

The numbers reinforced the hype. *Scandal* spawned 2.85 million tweets in season two, 25 percent more than the popular and live *American Idol*.[69] A mid-season episode generated nearly double the number of tweets as the anticipated premiere of *Game of Thrones*.[70] Once Nielsen's Twitter TV ratings were operational, *Scandal* was cited as one of the "most social" series on television. During the 2013–14 season, it consistently inspired more than 400,000 tweets that were seen by 2.5 million unique users and produced over 25 million impressions.[71] *Scandal*'s live ratings also continued upward, despite declines elsewhere on broadcast television. Near the end of season three in April 2014, *Scandal* was both the fastest-growing and highest-rated broadcast drama among viewers aged 18–49.[72]

The press linking *Scandal*'s ratings growth to the social chatter acted as a critical legitimating tool for both the series and for Twitter-oriented promotion. That live-tweeting mattered at all, let alone could help improve live viewership or convince a broadcast network to alter its promotional practice, signaled new definitions for industry success. At the 2013 upfront presentation to advertisers, ABC president Paul Lee called *Scandal* "a game-changing hit," touting its extensive social profile and engaged fanbase.[73] Lee's comments indicated that ABC believed that companies should place ads on *Scandal* not just because of its budding viewership but also because of its social footprint—and spend they did. Lee later praised Rhimes for bringing in nearly $300 million in advertising revenue, 5 percent of ABC's total figure.[74] As Nielsen worked to find a correlation between tweets and live viewership (and thus tweets and ad dollars), executives had already begun to act as if the relationship was self-evident.[75]

The increased media coverage of *Scandal* also focused on its appeal to Black Americans. By early 2013, *Scandal* was the most-watched drama among Black viewers, with more than 10 percent of Black households tuning in live.[76] A vital component of this appeal to Black women was the increased attention on the tumultuous relationship between Olivia and Fitz, which inspired the most euphoric live-tweeting among fans.[77] Rhimes had presented interracial romances before, but Washington's status as a rare Black lead and the salaciousness of an illicit relationship between her character and a white president made for an unprecedented and popular duo. Joseph argues that the "crossover approach" of Olivia—that she presented as a cunning, powerful woman without a racially coded backstory—enabled the romance with Fitz to appeal to viewers of all demographics.[78] As Anna Everett details, Rhimes's record for developing enchanting romances and encouraging fan passion online helped inspire all types of fans to "ride that emotional rollercoaster" and express their pleasure on the sub-hashtag #Olitz, a portmanteau of the character's names.[79] But as Warner argues, the chaotic relationship did offer specific appeal to Black women with a keen awareness of the historical realities of Black female representation: "Black women are rarely depicted as objects of desire on television . . . [they] are rarely allowed to be main characters in stories about choice, desire, and fantasy."[80] Both journalists and scholars highlighted *Scandal*'s brief references to race. A particularly notable moment saw Olivia tell Fitz that she felt "a little Sally Hemmings-Thomas Jefferson" about their relationship, prompting him to suggest later that she was "playing the race card" against him.[81]

While Rhimes continued to avoid discussing racial politics, *Scandal*'s nods to historical racism or interracial romance winked at audiences of color without undermining its multiculturalist perspective. To wit, National Public Radio reporter Gene Demby described the "deeply edifying, profoundly communal experience" among the Black audience at home and on Twitter: "*Scandal*'s adoring, hilarious, critical Greek chorus on Twitter—academics, pastors, teenagers, college kids, comedians, journalists, retirees, your auntie— feels like some kind of digitized concentration of Black American life."[82] Demby's celebration of Black conversations about the series overlapped with growing coverage of Black users on Twitter. Cultural critics had also begun to spotlight the influence of Black users on Twitter discourse. Writing for the blog *The Awl* in 2009, Choire Sicha called "Black Twitter" a "huge, organic" community that generated many of Twitter's trending topics and influential hashtags appropriated by white users.[83] In 2010, *Slate*'s Farhad Manjoo wrote an explanatory piece, "How Black People Use Twitter," which suggested how Black users "form tight clusters on the network—they follow one another more readily, they retweet each other more often."[84] Black critics like Kimberly C. Ellis and Shani O. Hilton quickly critiqued Manjoo for his view of Black Twitter as a racial and cultural monolith.[85] In his more recent analysis of Black Twitter, André Brock argues that Sicha and Manjoo did not fit the typical "exoticizing" of Black practices but also posits that the Manjoo column served as a "tipping point" in the wider (i.e., white) cultural imagination.[86]

The profile of Black Twitter grew along with the platform, from a panel celebrating the group's influence at tech and media conference South by Southwest to a glowing profile on *Today*.[87] Brock, Zizi Papacharissi, and Sarah Florini each propose that users of color employ distinctive language and semiotic codes online to denote group boundaries.[88] This process, known as "signifyin'," is similar to "code-switching," where people move in and out of linguistic contexts depending on the audience. On Black Twitter, Florini asserts, users employ "indirectness, doubleness, and wordplay" to perform membership of multiple identities.[89] Brock claims that while signifyin' can be understood "as a discursive, public performance of Black identity," the public nature of Twitter, complete with searchable hashtags, "invites an audience" from outsider groups.[90] Within the context of viewership, signifyin' enables a diverse discussion that can become more specific when Black users identify references that hold meaning for their subset of the audience. To an extent,

Scandal was structured to produce a passionate response from a segment of the active Black users at a time when Black users were especially active on Twitter.[91] Its brief moments of recognition of race or history were not dissimilar to shocking cliffhangers; they equally served to provoke reaction and conversation from fans on Twitter, particularly among a hyper-aware subset of fans. As Brock writes about Black Twitter in general, these reactions to *Scandal* could then be "repurposed" into future promotional campaigns on behalf of ABC and its corporate partners.[92] Thus, between Rhimes's post-racial ideology and the winking references to race, ABC tried to court Black viewers without alienating others. The network only expanded this crossover strategy, as well as the blend of conventional and modern promotional tactics, with the multi-platform-branded #TGIT campaign.

GET YOUR TWITTER THUMBS READY: BUILDING THE #TGIT UNIVERSE

Early promotion for the #TGIT program block personified ABC's synthesis of old and new tactics. While the materials underlined the emerging significance of hashtags and social media chatter, they also highlighted the hallmarks of Rhimes's auteur brand. The campaign pointedly also targeted both social media platforms and conventional broadcast and print media. A preview trailer circulated online and on television delivered a fast-edited montage of characters in stages of undress interposed by the appearance of textual exclamations like "OMG!!" The trailer sold new series *How to Get Away with Murder* as "from the executive producers of *Scandal*, the same people for putting the 'Oh no they didn't' in the Oval Office, comes your next TV obsession." The spot also celebrated that *Murder* would be another vehicle for a "brilliant & bad %*!" Black female lead, this time played by the renowned Viola Davis. The lively voiceover concluded, "With *Grey's*, *Scandal*, and *Murder*, you have the total trifecta of twisty, tweetable drama—all on one binge-worthy night. So get the red wine flowing and your Twitter thumbs ready—Thursday, September 25 is going to be awesome." As the final line hit, the trailer cut to bold-faced red text unveiling the hashtag/tagline, "#TGIT," with "Thank God It's Thursday" written in smaller font and parenthesis.

Immediately, then, #TGIT was celebrated for its sure-to-be "twisty" and "tweetable" nature, positioning storytelling conventions as equally as relevant as the tweets they produced. The instruction to the audience to prepare their "Twitter thumbs" and the recurrence of emotional bursts like OMG and exclamation points presumed that excited, "reactive" live-tweeting was standard for Rhimes fans.[93] Similarly, the presentation of #TGIT as a hashtag first and slogan second recognized that social media attention had become a central goal of ABC's promotional tactics. Still, the promo spotlighted steamy hookups and wild cliffhangers, hallmarks of Rhimes's storytelling long before social media. And Davis's presence in *Murder* was framed within an extended history of strong and diverse female characters on *Grey's* and *Scandal*.

Indeed, another early promotional item—an image that ABC turned into a print poster, billboard, digital "key art," and online banner ad—branded with the #TGIT moniker made this history more apparent. The artwork, illustrated with the same red, black, and white scheme used in the above trailer, positioned *Grey's* Ellen Pompeo, Washington, and Davis from left to right, indicating both the historical expansion of Rhimes's brand and their programmed flow on the schedule. The highly edited image placed the women in the same visual space to convey the idea that they, and series, were part of a larger shared world.

#TGIT extended the simulated liveness and flow first constructed by *Scandal*. But while this promotional blitz ostensibly aimed to convince people to watch ABC live on Thursdays, the ongoing cross-pollination of #TGIT figures—on social media during organized live-tweeting, in commercials promoting the program block, and later in two of the series—established the shared universe of Shondaland. This evolution further proved that networks find ways to incorporate more genuine forms of audience engagement into conventional frameworks.

In the lead-up to *Murder* and #TGIT's debut, industry reporting predictably focused on the diversity of the Shondaland universe and the importance of Twitter, particularly the multi-platform attention secured by *Scandal*'s #WhoShotFitz initiative. Rhimes again downplayed the "unmatched in TV history" programming block and the role of race in her growing empire.[94] As she told the *Los Angeles Times*, "everybody thinks it's a bigger deal than I do.... But I'm not really thinking about, 'Oh, I'm the Thursday queen.' Uh, no. I'm thinking, '*Grey's* has to be good, *Scandal* has to be good, and *Murder* has to be damn good.'"[95] Rhimes also upheld the color-blind casting approach, even as the press praised the presence of another Black female lead on *Murder*. "Why did it take somebody black to talk about being black?" Rhimes said to the AP. "There is no token system. We're gonna cast the best actor for the part. And then our cast makes it feel real and true—they elevate everything."[96] For her part, Washington embraced the cultural significance placed upon #TGIT by the press—but also situated that significance in the realm of social media. As she said to the *Los Angeles Times*, "It's pretty extraordinary.

And important. And fabulous. And to be celebrated. And I don't think Twitter is ready."[97]

#WhoShotFitz, meanwhile, was cited as the inspiration for ABC's full-scale embrace of social media promotion. Goldwyn admitted that, while before #WhoShotFitz, the actors "were hype on a grassroots level," the corporate barrage for the hashtag helped live-tweeting catch on "like wildfire."[98] In fact, Twitter had become so crucial to ABC that network officials asked Rhimes and Peter Nowalk, the creator of *Murder*, if they had considered the promotional challenges of translating the program's long title into an easily shareable hashtag.[99] Meanwhile, Rhimes and Shondaland were credited with exposing the value of live-tweeting to the entire industry. According to Twitter executive Anjali Midha, *Scandal* and ABC were "driving the kind of behavior that advertisers and networks really care about."[100] As Nielsen's Twitter TV Ratings and reports about the increases in multi-screen use circulated in the press, the influence of live-tweeting was no longer speculative or seen as a fun gimmick.[101] Executives, actors, and corporate partners agreed about its value to the viewing experience—and the bottom line.

But despite the focus on the novelty of Twitter in the press, ABC again emulated existing industry maneuvers with #TGIT, most notably in the planned flow of the programming block. Ellis argues that scheduling decisions represent "the point where the activity of the past and the hopes of the present become the strategy of the future."[102] With #TGIT, ABC sold viewers not only a continuation of Rhimes's twisty and tweetable series but also a remix of the network's history. Beginning in 1989, ABC branded its popular Friday family sitcom block with the TGIF slogan. As one of the most notable themed nights, TGIF used cross-promotional segments with actors introducing the night's content and later commenting on it between episodes.[103] TGIF actors appeared "live" on their respective sets, often sitting on a prop couch near a television, to directly address the audience to implore them to keep watching.[104] TGIF was designed as a synchronous expression of television scheduling, flow, and liveness. Its success inspired other themed nights on ABC's schedule, as well as on the schedules of other networks like NBC, which famously began referring to its Thursday series as "Must-See TV" in 1993. With #TGIT, ABC expanded the interstitials from the television screen to the second screen. TGIF presented an artificial connection between actors and fans, as the former *spoke at* the latter through the television to try to carry viewers through the flow of the program block. #TGIT, conversely, proposed an immediate conversation, with performers and fans potentially *speaking to* one another online. Yet, the structured flow of ABC's schedule similarly hoped to accentuate the tweetable moments of #TGIT and motivate fans to watch and discuss live.

#TGIT appeared in tweets of all types. Indeed, in the week before the first night of #TGIT, notable Shondaland accounts produced a similar stream of tweets, mixing promotional photos, videos, links, red-carpet snapshots, and previews:

> @KatieQLowes: The Scandal premiere day is now close enough to see on a 10 day weather forecast. #TheLittleThings (September 15, 2014)
>
> @shondaland: 9 days. Like a kid waiting for Christmas... #GreysAnatomy #Scandal #HowToGetAwayWithMurder #TGIT #shondalandtv (September 16, 2014)
>
> @violadavis: 224 hours, 36 minutes, and 30 seconds until @HowToGetAwayABC #TGIT #ShondalandTV #HTGAWM (September 16, 2014)

These tweets fostered anticipation for #TGIT's premiere and directed interested fans to additional touchpoints. Yet, they also denoted usual forms of promotion. That they were shared on Twitter only recontextualized the delivery system, not the nature of the tweets. These tweets functioned as what Jonathan Gray calls "entryway" hype, which holds "considerable power to direct our initial interpretations."[105] I would add that the influence of entryway material is more significant when shared by stars who perform an authenticity to drive interest in their career.[106] Through the power of linking, hashtagging, and "@ing," these tweets continued to push the existence of a collective #TGIT universe that was established in conventional promotional materials (posters, magazine covers, live events) featuring Pompeo, Washington, and Davis.[107]

Though the growth of Twitter increased the preseason promotion's profile, it also ensured that it would seem significant when actors from different #TGIT franchises live-tweeted at one another during their respective episodes. Karoline Andrea Ihlebaek et al. argue that modern programming and scheduling is best identified as a universe approach that accommodates a range of platforms and audience interest levels.[108] Reference to the media universe recalls the shared Marvel Cinematic Universe, where the studio's superheroes have appeared together in increasingly more films and television series as part of a larger storyworld.[109] But universes like Shondaland do not require a common *fictional* diegetic reality when their key figures can utilize social media to mingle publicly and tease fans about eventual crossovers within those fictional realities. The Shondaland universe was affirmed during #TGIT's debut on September 25. #TGIT appeared on-screen during the episodes while Rhimes and actors tweeted throughout the night, priming viewers to stick with all three hours to get the full #TGIT experience. Davis, the newest lead participant, sent several tweets reminding her followers about *Murder*'s premiere, including one with the familiar red promotional image of her and her universe co-stars:

> @violadavis: Going to #Shondaland for 3 straight hours. #GreysAnatomy, #Scandal & #HowToGetAwayWithMurder starting now! #TGIT

Then, as the episode was about to conclude, Davis tweeted again:

> @violadavis: 2 minutes left . . . Do. Not. Blink. #HowToGetAwayWithMurder #TGIT

Davis's phrasing of "Going to #Shondaland" underlined the enterprise as both worthy of a hashtag and a genuine destination—one that required an extensive but ultimately worthwhile time commitment to visit. The urgency of the second tweet promised that *Murder*, like its #TGIT brethren, would be so twistable and tweetable that viewers should not blink lest they miss something important on either screen. Davis first tried to escort viewers from the second screen to the television, signaling that she would be joining them on the journey to Shondaland. She later worked to hold their attention through the climax of the episode and the first night of #TGIT. Both tweets assumed a live, shared, and multi-platform experience, down to the exact minute, that kept everyone attuned to #TGIT on all screens.

Washington remained a passionate tweeter willing to chat with her co-stars, praise her new Thursday night friends, and celebrate the engagement of the fans on Twitter:

> @kerrywashington: "@KatieQLowes: Breathe, you guys, breathe. #TGIT" I CAN'T!!!!

> @kerrywashington: YES!!!!!! Viola!!!!!! YES!!!!!! #TGIT #HTGAWM @violadavis

> @kerrywashington: #Scandal fans are THE BEST FANS EVER. Thank u for watching. Thank u for tweeting. Thank u for being. #TGIT It's goooooooood to be back!

Washington positioned herself as a diehard fan of all things #TGIT. She previously encouraged fans to try to control their breathing; now, she could not. Washington was so excited for Davis and *Murder* that all she could tweet was "YES." Her tweets unveil the base emotional response at the center of #TGIT live-tweeting. The series stirred all-caps exclamations, even from their stars. Rhimes, meanwhile, remained the most active tweeter, offering anecdotes throughout the East and West Coast premieres of each series. Her tweets ran the gamut from fan appreciation to emotional reflections on specific scenes to promotions of the Shondaland shared universe:

> @shondarhimes: My favorite thing about tweeting with my #Shondaland family is @EllenPompeo who is not talking because she is busy eating. So cute. #TGIT

> @shondarhimes: This scene kills me. Goodbye Harrison. #scandal #TGIT

> @shondarhimes: Two down, one to go! Put down the red wine. Pick up the vodka. And get ready to meet Annalise Keating! #HowToGetAwayWithMurder #TGIT

> @shondarhimes: Tweeting with my #Shondaland family!! Me and @TheRealKevinMcKidd #TGIT

> @shondarhimes: We have been tweeting for almost five hours and we are having a blast! #scandal

Rhimes's tweets were both promotional and personal. Like Washington, she performed an appreciation for fan interest in her work. And like Davis, she reminded followers of the next episode to come on the schedule. But Rhimes also brought the Shondaland universe to life by tweeting her observations of Pompeo eating as they watched together and later sharing photos of *Grey's* performer Kevin McKidd and the *Scandal* cast lounging on a couch, phones in hand. These posts assured fans that actors spent time together off-screen, sent their own tweets, and had a "blast" doing it. The use of the words "we," "us," and "family" fostered a sense of community among Rhimes, the actors, and her fans. As the face of Shondaland, Rhimes operated like a host, guiding fans through the proper emotional responses, interpretations, behind-the-scenes tidbits, and community building, ensuring everyone was "having a blast."

Naturally, ABC integrated classic cross-promotion into the #TGIT chatter. Shortly before the September 25 premiere, each of the series Twitter accounts publicized a cross-branded sweepstakes contest with the upcoming Twentieth Century Fox film *Gone Girl*. On September 22, @ScandalABC shared the rules for the "#TGIT LIVE #TWEEPSTAKES presented by Gone Girl," instructing fans to live-tweet using both hashtags and follow the series on Twitter. Participating users would then be selected at random to win a "#TGIT Survival Kit" featuring #TGIT wine glasses, chocolate, coasters, "Olivia Popcorn," a popcorn bowl, a T-shirt, and a copy of the *Gone Girl* novel. The images featured the red, white, and black color scheme, links to each series' Twitter account, repeated use of #TGIT, #TWEEPSTAKES, and LIVE. On premiere day, many users included #TWEEPSTAKES in their excited tweets about #TGIT:

@darkandtwisty: @shondarhimes SHONDA I NEED THIS #TGIT #TWEEPSTAKES

@Kim_Rendino: Woohoo!! Almost time to kick the hubby out to watch football and get MY Thursday night started!!! #TGIT #TWEEPSTAKES

@janiewalla: Got tequila, wine, popcorn . . . Could probably use a survival kit . . . #TGIT #TWEEPSTAKES #shondaland

@BrLittle2: Phone charged. NOW I'm officially ready! #TGIT #Tweepstakes #ScandalThursday #HowToGetAwayWithMurder @Scandal @HowToGetAwayABC

Free contests, from mail-in competitions to radio giveaways, are commonplace. By bringing them to Twitter, ABC and Twentieth Century Fox put a faux-participatory spin on the "content-promotion hybrid" strategy, with conversation at the center.[110] The simple use of a hashtag in a reactionary tweet increased attention for Rhimes's projects and the film. This maneuver stressed the importance of the live #TGIT experience and the structured flow of ABC's Thursday schedule. Fans could talk to stars *and* win merchandise, but they had to tweet throughout the night to increase their chances of both outcomes.

Building the Shondaland and #TGIT universe across platforms paid off immediately for ABC. With Rhimes and actors tweeting alongside fans, the #TGIT premieres delivered high ratings on television and Twitter. *Grey's*

Promotional Twitter image for the #TGIT-*Gone Girl* sweepstakes encouraging fans to "live-tweet tonight with #TGIT and #tweepstakes" in hopes of winning a package of branded apparel and snacks.

and *Scandal* returned with larger audiences than the previous season, while *Murder* was the highest-rated new series of the season. #TGIT gave ABC its highest-rated Thursday night in five years, and all three series were popular among young female and Black viewers.[111] Twitter told a similar story, where each series finished in the top six most-tweeted-about series of the week.[112] ABC boasted about the success of #TGIT in a promo aimed at securing even more viewers. Styled like the preseason promotional video, the twenty-second clip celebrated how, "with over 35 million viewers and over 1 million tweets, #TGIT on ABC is a total TV phenomenon." The spot suggested that television viewership and social footprint (via the enthusiastic fan activity) were equal factors in #TGIT's phenomenon stature. #TGIT sustained viewership and Twitter activity through the fall, with the three series remaining in the top 15 for both Nielsen viewership and Twitter TV Ratings.[113] Across the fall episodes, the series amassed 5.2 million total tweets, with #TGIT mentioned in 836,680 of those tweets.[114]

The success of #TGIT and another Shondaland drama brought more coverage about the live-tweeting strategy. "#TGIT led the charge," said Lara Cohen, Twitter director of talent. "The shows definitely set the gold standard."[115] ABC executives likewise declared that *Murder* and #TGIT convinced more viewers to watch live. "#TGIT showed that people can come to one show and stay the whole night if there is something that has a true natural flow," said executive Marla Provencio.[116] ABC found that *Scandal* was mentioned in 6 percent of the tweets about *Murder*, implying some obligation to the flow on Twitter as well.[117] Here, network and Twitter executives worked to prove that tweets aided—and not disrupted—television flow.

By early 2015, live-tweeting and #TGIT were ensconced in the minds of viewers and industry representatives. ABC's promotion for #TGIT's post-holiday return poked fun at the block's popularity. Across multiple ads, Shondaland actors solemnly explained the symptoms of "#TGIT Withdrawal Disorder." The spots, framed like hokey ads for prescription medication, detailed that anyone missing their favorite dramas was not suffering alone. As the cast of *Murder* explained:

> Having #TGIT off the air hurts everyone. Me, you, us. Research shows that #TGITWD is caused by an OMG deficiency in the brain. #TGIT works to correct this deficiency. Simple activities like talking, tweeting, or rewatching with friends can help until #TGIT returns January 29.

Having learned that #TGIT would not return until late January, *Murder*'s Matt McGorry dropped to his knees in sadness and, blurring the lines between actor and character, called out for Olivia Pope. As McGorry shouted, "Why?!!" other Shondalanders looked into the camera to urge the audience to "know the signs. Know the return date. And on January 29, no more waiting." The spot finished with instructions: "Share your #TGITWithdrawal stories with videos, GIFs, and pictures," all formatted with the familiar red/white color scheme. Another spot saw *Scandal*'s Lowes and Scott Foley interrupting one another with gradually parodic acts of #TGITWithdrawal, including a black-and-white shot of a depressed Lowes labeled "DRAMATIZATION." The duo explained that while viewers rewatched #TGIT, they should call their doctors if they experience "outbursts of OMGs . . . uncontrollable tweeting . . . heart palpitations, eyes popping out of your head, hot flashes, cold sweats, dizziness, or a good ol' fashioned fainting spell." They then instructed viewers to "ask your friends if #TGIT is right for you. And if they say no? You should probably just find other friends."

The #TGITWithdrawal campaign encapsulated years of work by ABC. The commercials referred to the significant Rhimes storytelling conventions, the strong audience response, the role of social media engagement, and the importance of live-tweeting. They also further ratified the existence of a shared universe, with actors stressing to fans that they were part of a community of "millions" in Shondaland. That #TGITWithdrawal self-reflexively joked about these conventions of the Shondaland universe signaled that fans had accepted the conventions, even to the point of winking ridicule. Here, #TGIT programming was so influential that fans were sick without it; however, the list of symptoms also implied that things such as excessive tweeting were part of the weekly #TGIT experience. #TGIT was both the illness and the cure, thus operating as a dominant force within fans' lives. Of course, fans had no problem playing up this idea on Twitter, responding with the requisite hashtags and self-reflexive humor about their fandom:

> @Olitz4Eva: Sitting here waiting for Scandal to come back on . . . Like BRUUUHHHH I NEED OLITZ #TGITWithdrawal #TGITWD (January 23, 2015)

> @Rachelree1023: RE: My last tweet. Why is #TearsofGreys a hashtag? WHY SHONDA, WHY??? So now along with my #TGITWithdrawal, I've got #TGITAnxiety (January 23, 2015)

> @MikkiWesley: #TGITWithdrawal #TGITitis KILLING ME!!! (January 24, 2015)

As with most jokes, the campaign and the tweets indicated that the withdrawal was, to a degree, real. The tweets display that fans quickly riffed on the illness framing to generate new hashtags that still emphasized their legitimate anticipation for #TGIT's return. The #TGITWithdrawal campaign showed that Rhimes, Washington, and the entire Shondaland crew helped build a community that fans wanted to be part of, not only live on Thursdays but throughout the week and across platforms. The campaign also showed that ABC capitalized on the goodwill of the community via scheduling and promotional maneuvers that used or referenced social media but indeed served to accentuate the power of liveness and flow.

CONCLUSION: #TGIT CROSSES OVER, AND BURNS OUT

The press coverage of #TGIT as a revolutionary event helped to articulate the relevance of Twitter to larger constituencies at home and within the media industries. Although Washington is credited for the expansion of *Scandal* live-tweeting, #AskScandal, and #TGIT, the success was also the result of an amalgamation of storytelling conventions, planned scheduling, celebrity, and branding tactics. Twitter chatter about *Scandal* did help build its popularity and convinced ABC to build an entire night of Shondaland programming, but *Scandal* had to generate the big moments that made people want to watch and live-tweet in the first place. While fans responded to Washington's early live-tweeting, they might not have without Rhimes's enduring willingness to engage online. And they would have been even less inclined to watch *Scandal* if it were not for Rhimes's distinctive voice as a writer—or *Scandal*'s placement on the ABC schedule directly after *Grey's*.

To a degree, #TGIT confirmed the remaining power of broadcasting in the face of mounting pressure from streaming portals. Twitter provides a platform for people to reify and deepen bonds as part of a communal experience, heralded by hashtags, trending topics, and memes. But those markers of community are often driven by an inciting reference point found elsewhere, something like a dramatic cliffhanger or shocking hookup in a popular television series. The planned nature of the television schedule can create and maintain anticipation for a night's worth of inciting incidents, even as viewers split their attention across multiple screens during those crucial moments. *Scandal* and later all of #TGIT showed that a strategically cooperative flow—where events on television are synchronized to content or chatter on a social platform, or vice versa—can generate a constructed collective experience to convince audiences to watch live. Twitter chatter expanded the feelings of a shared experience, but the aura of liveness remained central to the experience as well.

It was unsurprising, then, to see ABC leverage fan energy for positive public relations efforts and then turn that energy toward far more conventional promotional strategies that also included hashtags for targeted bursts of viral promotion. Though ostensibly novel with their focus on hashtags or social impressions, these initiatives still pushed audiences to watch in the most traditional way possible: live without skipping any of the commercials. Within the context of new metrics for audience measurement, the industry's fervent embrace of #TGIT as a model for Social TV strategy held added meaning. The success of #TGIT suggested that live-tweeting could impact conventional ratings and build online advertising revenue. But it also allowed ABC to continue to celebrate the diversity of Shondaland and its allure to nonwhite audiences in a new realm like Twitter, without weakening Rhimes's color-blind perspective. Indeed, as Mary Ingram-Waters and Leslie Balderas argue, #TGIT flattened the sophisticated conversations and community-building among Black women on Twitter into marketable reactions like "OMG."[118] Thus, #TGIT's position as a hyper-modern campaign with politically progressive outcomes elided the network's never-ending attempts to govern audience response.

The incorporation of social conversation into customary promotional methods is no more apparent than in the evolution of the Shondaland universe. Chatter among colleagues online undoubtedly aided in attracting viewers to ABC's Thursday schedule, but the network progressively pushed the actors together more across platforms to sell the idea of a cohesive Rhimes universe to the point of parodying its impact on fans in the #TGITWithdrawal commercials. Although universe-building has its purchase in Hollywood and with fans, ABC's subsequent attempts to expand the universe unveiled some of the approach's limitations. During the 2015–16 season, ABC unveiled the newest addition to the #TGIT lineup with Rhimes's latest female-fronted series, *The Catch*. To mark the occasion, ABC produced a round of familiar ads that integrated *The Catch* star Mireille Enos into Shondaland. Veteran #TGITers welcomed Enos into the fold with a salute of a red wine glass, a signifier of *Scandal* and the Thursday programming block. But *The Catch* was canceled after brief two seasons, indicating that even the combination of Rhimes's brand, the scheduling halo of #TGIT, and familiar promotional markers did not guarantee success.

Soon after, ABC announced that *Scandal* would conclude in April 2018 and thereby end the original #TGIT programming block.[119] However, the series did not leave the air without at least one more major twistable and tweetable moment: a direct crossover with *Murder* that brought Washington's Olivia and Davis's Annalise together on-screen beyond a quick-cut commercial or publicity photo shoot.[120] ABC promoted the event with a simple hashtag: #TGITCrossover. In some ways, the *Scandal-Murder* crossover finalized ABC's co-opting of the early social media enthusiasm demonstrated by Rhimes and then later Washington. This appropriation served as another kind of new flow that sketched the Shondaland universe from social platforms to promotional campaigns to the storyworlds of two series.

But in other ways, the crossover was a final bow for #TGIT as it was conceived initially—and for Rhimes as the face of ABC. In 2017, she partnered with Hearst Magazines and corporate sponsors like Dove to launch Shondaland.com, a branded lifestyle website pitched as an "empowering and inclusive platform for women." Featuring a combination of first-person essays, style tips, profiles of "real women," and reported pieces about feminism and racial politics, Shondaland.com enabled Rhimes to convert years of tweeting and fan engagement into a new endeavor that expanded her branded universe once more.[121] Soon after, Rhimes revealed that she would be leaving ABC Studios and signed a $100 million production deal with Netflix. For Rhimes, the move was yet another sign that she had become one of the industry's most influential voices. For broadcast television, her ABC departure was seen as another knock against the rigidity of the network system in comparison to the creative freedom now granted by streaming video companies.[122]

That Rhimes's departure from ABC would be framed as the end of an era is unsurprising. A lot changed between the *Grey's* premiere in 2005 and the *Scandal* premiere in 2012; even more changed between 2012 and 2018. The emergence of Netflix, Amazon, and Hulu as willing spenders created a new arms race in the development of scripted television. Broadcast networks have often been left out of the arms race altogether. This chapter illustrates that, at least for a time, broadcasters like ABC adapted just enough to remain essential industry players. But the surprising plot twists that made *Scandal* and *Murder* so tweetable also aided in their creative downfalls, and in burning out the fans who were once so ready to share their reactions with others. Likewise, the novelty of campaigns like #TGIT wore off as other networks pushed corporatized hashtag initiatives without as much organic buy-in from key cast and crew constituencies. These developments helped shrink the palpable excitement surrounding the ephemeral experience of live television and live-tweeting in a few years.

The following chapters detail case studies from other networks and companies that frame their technology, programs, or multi-screen experiences as more innovative than what network television has to offer, or beyond the purview of Nielsen's evolving ratings. But the sincere triumphs of *Scandal* and #TGIT were as much a product of liveness and flow as live-tweeting and hashtags. Other companies would try and mostly fail to capture the kind of attention achieved by ABC in the mid-2010s.

Chapter 2

IMMERSE YOURSELF DEEPER

Building AMC's Multi-Screen Storyworld

On May 27, 2016, cable network AMC released an online audio commentary for the pilot episode of its latest buzzworthy series, *Preacher*. Based on Garth Ennis and Steve Dillon's popular 1990s supernatural comic series, *Preacher* traveled a winding road to live-action adaptation before finally landing in the hands of Seth Rogen and Evan Goldberg, one of Hollywood's most prolific comedic duos. Though initial reviews and Nielsen ratings for *Preacher* were promising, AMC used the Memorial Day weekend to reach fans of the comics who may have been skeptical of an adaptation under the auspices of Rogen and Goldberg. The commentary gave the co-creators a platform to prove their fanboy bona fides, with Rogen stressing, "We've loved *Preacher* forever. We read it when it came out in the '90s, and we've been trying to make it for around ten years, basically."[1]

The *Preacher* directors' commentary was just one of AMC's attempts to sell viewers on bonus content related to tentpole series like *Preacher*, *The Walking Dead*, and *Better Call Saul*. This content includes 1) behind-the-scenes footage and cast interviews offering insider anecdotes about the production process; 2) short webisodes and deleted scenes delivering extensions to the television narrative; 3) "interactive" tours of sets and maps of important locations lending credence to the series as sprawling storyworlds; 4) blog posts informing fans with relevant news; 5) competitive quizzes and trivia contests labeled as "ultimate fan games"; and 6) forums and chat rooms prompting fans to "join the conversation" during live episodes.

But a few years prior, the centerpiece of AMC's digital strategy was Story Sync, a real-time, "audio-watermarking" second-screen experience professing to empower viewers to "interact with shows while they air." First introduced in 2012, Story Sync played alongside the live broadcast of new episodes on the second screen. As the live episode aired on television, Story

Sync dispensed narrative context through flashbacks to prior sequences or similar events from a program's source material (e.g., *The Walking Dead* comic). Story Sync also asked viewers to engage in playful predictions and trivia contests connected to the on-screen action or the thematic concerns of the series. Crucially, while portions of Story Sync content were accessible to all viewers at all times, the full "interactive experience" was only available to those willing to watch episodes—as well as commercials—during the live broadcast. Promising viewers that they could "immerse" themselves "deeper" into AMC dramas, Story Sync consolidated familiar ancillary content into an interactive and live package.[2]

AMC was not alone in efforts to drive viewers to watch programs live, visit a website, or download an app. In response to changing viewer habits, networks have posted full episodes and bonus material online for more than a decade. AMC and cable competitors also used Twitter to circulate promotional material and promise forms of real-time participation. Like ABC's #TGIT campaign, AMC courted live viewer engagement with a mix of targeted hashtags and on-screen prompts. AMC also utilized the aftershow format where personality Chris Hardwick recaps episodes for a studio audience, interviews performers, and takes questions from Twitter. *Talking Dead* signaled AMC's efforts to integrate Twitter chatter into live episodes. But the budding focus on live-tweeting did not dissuade networks like AMC from testing multi-screen products built on what they know best: programming. By framing its series as worthy of streams of bonus content, AMC challenged the prevailing assumptions about Social TV and supported the claim embedded in a product name like Story Sync: viewers should interact *with content* rather than interact *with one another*.

Drawing from an analysis of the Story Sync platform, as well as press coverage of and fan response to the product, this chapter investigates how AMC constructed the second screen as a space for the celebration of its acclaimed dramas. With Story Sync as the central case study, I demonstrate how AMC situated specific serialized series as sprawling storyworlds that *required* an extra screen's worth of content to capture it all. Story Sync's intertextual nods to the source material, prior narrative events, and even US history suggested that *Walking Dead* and *Better Call Saul* were deep, "drillable" texts.[3] These nods were aided in real time by Story Sync's circulation of familiar commentary from the cast and crew underlining the care and skill needed to make these series. But despite the alleged scope and scale of these multi-screen storyworlds, Story Sync also promoted an interactive, spirited, and gamified live experience that moved at the speed of an hourlong serial. Presenting viewers with predictive trivia contests and provocative, yet playful, moral quandaries, Story Sync strived to fit within the rhythms of social chatter and mobile gaming.

The multi-screen product celebrated fan expertise but also required participants to engage and align with series' core themes, what I call *synchronized reiteration*. Story Sync's vision of Social TV used the second screen to stress to viewers what ancillary information they needed to understand better. Fandom on Story Sync reflected what Suzanne Scott refers to as "competency in masculinized modes of fan engagement": accumulation of knowledge in the form of trivia, behind-the-scenes tidbits, and approved interpretations.[4] As Taylor Nygaard and Jorie Lagerwey argue, discourses about immersive or complex narratives have functioned to separate male-centric cable series from the historically feminized broadcast networks.[5] This is part of the legitimation of television in the twenty-first century, but it also speaks to a broader movement in media franchising described by Derek Johnson that aims to elevate "feminized" serial narratives with masculinized industrial strategies.[6] With its focus on immersive storyworlds and masculinized competencies, Story Sync reinforced conventional gendered notions of "important" television, and what type of programming is allowed to personify depth or sprawl across multiple platforms. Story Sync denotes another example of the media industries' attempts to, as John T. Caldwell and Barbara Klinger respectively contend, "repurpose" promotional content onto new platforms.[7] But, as illustrated throughout the Social TV era, repurposing applies to strategy as much as content. The temporal specificity of Story Sync—that viewers had to tune in live to get the "true" experience—tried to recapture the ephemerality of television in the face of significant time- and place-shifting. Yet, Story Sync also borrowed from the DVD special feature and multi-platform and transmedia storytelling, antecedent approaches from prior generations of digital media disruption. Thus, Story Sync promised to expand a franchise's storyworld *outward* in perpetuity, underlining linkages across platforms, formats, and histories, by asking fans to drill *inward*, into its story and themes. The competing components of Story Sync—between drillable multi-screen text and fun live game; between informative DVD feature and interactive social experience; and between ephemerality and permanence—illustrate the hurdles in importing past strategies into the present.

The inherent tensions of Story Sync influenced its short lifespan. In 2013, Story Sync was nominated for the Outstanding Interactive Program award at the Creative Arts Emmys, the premier awards for below-the-line artists.[8] In 2014, AMC executive Mac McKean reported that weekly usage of the Story Sync platform "favorably compare[d]" to the number of people (between

350,000–500,000) who were tweeting about *Walking Dead* at that time, and, that throughout one season, unique user figures ran "well north of 1 million."[9] However, while the scale of Twitter activity linked to a popular series like *Walking Dead* continued to grow—1.3 million tweets were sent about the 2015 season premiere alone—Story Sync did not see the same expansion.[10] Despite the connection to popular franchises and an initial wave of press, these immersive second-screen experiences failed to sustain interest. Similar products from HBO and Showtime quickly disappeared, while AMC slowly phased out Story Sync until its cancellation in 2017.[11] This failure indicates that viewers, empowered by industry rhetoric about choice, did not embrace a second-screen experience with such noticeable corporate structure, and such disregard for the *social* elements of Social TV.

In detailing the brief existence of Story Sync, this chapter extends the conversation about ephemeral media, which Paul Grainge describes as "evanescent, transient, and brief . . . short-lived . . . concerned with the peripheral and throwaway."[12] Both the temporality and life cycle of Story Sync embodied the spirit of ephemerality. Excavating material from Story Sync can be difficult; only portions ever appeared online, unsynchronized from the live experience, and that small pool of content has been removed as AMC programs have ended. As AMC now promotes other "interactive features" like virtual reality simulations or special emojis, Story Sync has become part of what Amelie Hastie refers to as the "discarded or temporary forms and sites" of media culture.[13] Cases like Story Sync signal how quickly media companies are willing to eliminate the material related to botched digital experiments. If, as Klinger contends, ephemeral content fits "seamlessly into both the surfing mentality that defines media experience and the multitasking sensibility that pervades computer culture," it is no surprise that media companies abandon it at the sign of failure.[14] Corporations assume that consumers only have so much attention to give to the array of entertainment choices and will not care or remember that most ephemera disappear. Story Sync points to the contested nature of contemporary ephemeral media. When identified as a potential boon, ephemeral content is overemphasized and its influence is overplayed. When considered failed or passé, it is just as easily discarded altogether.

Placing Story Sync within a history of DVD features, multi-platform content, and transmedia storytelling, this chapter is also concerned with textuality: texts, paratexts, and intertexts. The digital environment has expanded the circulation of what most would consider "primary" texts: films, television series, books, games, and so on. Yet, the spread of devices with large storage capacity and high-speed internet accessibility has also aided in the repurposing of paratexts, which prepare—or hype, inform, and structure—the consumption of a primary text.[15] Content that once made up DVD bonus discs now populates websites, social channels, and, in the case of Story Sync, multi-screen experiences. Jonathan Gray argues that paratexts are "not simply add-ons, spin-offs, and also-rans: they create texts, they manage them, and they fill with many of the meanings that we associate with them."[16] As Gray notes, "entryway" paratexts like trailers attempt to prime audiences for specific interpretations of a film *before* they see it. But Story Sync content compressed the common temporal scheduling of paratexts through a *real-time* affirmation of the dominant themes and ideologies of its series.[17]

Part of the intended interpretive power of Story Sync came from its deployment of intertextuality, or the references, allusions, and connections to other texts. Mikhail Iampolski argues that, while intertextual references attempt to "place what you see alongside what you know, alongside what was already been . . . only the viewer or reader can unite the text, using his cultural memory to make it one."[18] In contrast, Michael Riffaterre claims that intertextuality is a method used to "guarantee" that readers make the proper connections between and readings of texts.[19] Authors, producers, or marketing teams cannot guarantee any interpretation from the audience, but they can construct circumstances where intertextual references stress the desired reading.[20] This is, again, where "official" paratexts come in handy, as they push what Melissa Aronczyk calls "an authoritative aura" and "assumption of responsibility" for meanings.[21] In presenting what Hye Jin Lee and Mark Andrejevic refer to as a multi-screen "digital enclosure," where paratexts guided viewers to precise franchise touchpoints, intertextual references, and authoritative meanings, Story Sync illustrated Julia Kristeva's assertion that textuality is "an intersection of textual surfaces."[22]

Below, I review Story Sync's ancestors in repurposing DVD bonus features and transmedia initiatives and their role in regulating viewer interpretations. Next, I trace the creation of Story Sync and its place within the then-emergent Social TV sphere, paying special attention to how AMC executives and promotional material framed the product both as an immersive tool for fan knowledge and auxiliary space for quick moments of gamification. I then analyze archived Story Sync interfaces and content to underline how AMC aimed to manage viewer experiences and keep them away from a more populated social platform like Twitter. Finally, I survey audience response to Story Sync on Reddit, where users negotiated its utility to their collective speculation about narrative developments. The confusion with Story Sync among this small selection of users highlights how even the most-hyped "official" paratextual enterprises can backfire, or simply peter out, with target audiences.

FROM THE SHELF TO THE SECOND SCREEN: PRECURSORS TO STORY SYNC

Not that long ago, the DVD was the most advanced home distribution technology. Introduced to US consumers in the late 1990s, DVD players quickly began to sell at the rate of 350,000 per quarter.[23] Though part of a long lineage of home entertainment, the DVD was also highly successful in normalizing an informed consumption through special editions and bonus features.[24] Klinger details how feature-laden collector's editions first served as niche artifacts targeted to "film buffs and academics," but they caught on with more general consumers.[25] Audio commentaries and behind-the-scenes production diaries, colloquially referred to as "film school in a box," offer unmatched access to directors and sets.[26] These features follow a project through the production process, from casting and location scouting in preproduction to sound design and effects work in postproduction. Mark Parker and Deborah Parker assert that, by providing an inside view of production, DVD extras sell "heightened attention to intricacies of intention as it plays out."[27] The DVD also offers consumers what Paul McDonald calls "multiple axes of control," including bonus features, menus, subtitles, and language options.[28] For viewers of television series on DVD, control means evading scheduled flow where episodes are broken up by commercials. DVD box sets moved television into the realm of "tangible" and "symbolically bounded" objects like novels or paintings, giving consumers ownership of content that was once more transient.[29] In promising more control at home and behind-the-scenes tidbits, the DVD was situated as an interactive format with a backdoor to Hollywood wonders.

Nonetheless, DVD features generate a positive portrayal of industry professionals and production cultures. As Klinger writes, "Far from demystifying the production process, these revelations produce a sense of the film industry's magisterial control of appearances."[30] They often center on the director to perpetuate the ideology of singular authorship for promotional purposes. Below-the-line artists, meanwhile, act as support for the auteur's voice. Commentaries make this contrast of artistic agency more apparent with directors speaking about their vision despite the collaborative nature of filmmaking. Robert Brookey and Robert Westerfelhaus write that special features frame participants "as having privileged insights regarding a film's meaning and purpose, and as such, they are used to articulate a 'proper' (i.e., sanctioned) interpretation."[31] Thus, while DVD features facilitate critical engagement with film, they also "deliberately add new layers" to audience interpretations that often match Hollywood's self-mythologizing."[32] Likewise,

DVD features still restrict the control over the viewing experience, and only encouraged more attentive watching and rewatching. Greg M. Smith claims that increased control is a "fantasy" concocted by manufacturers and studios that presumed that older forms of media are consumed passively.[33] In this regard, the DVD was similar to prior iterations of home video: more control led to more consumption.

Perhaps most important, the DVD epitomizes Hollywood's desire to reuse content in new channels and for new audiences. For John T. Caldwell, DVD features emulate a core Hollywood promotional mechanism known as the "press pack" or "electronic press kit" (EPK), which were circulated among the press to aid prerelease coverage of films or television series.[34] Studios simply shifted the flow for this material, refashioning interviews and production diaries as value-added content for home video releases.[35] These features denoted what Caldwell calls "textual dispersal," allowing the media industries' to navigate new consumption habits through low-risk practices.[36]

For a certain segment of consumers, special features create a precarious scenario. On the one hand, the attentive viewing enabled by DVDs allowed these viewers to embrace their status as targeted "insiders." As Matt Hills argues, production tidbits and commentaries convince audiences "to feel that they are part of an inner circle of knowledgeable consumers."[37] Insiders feel, Klinger says, as if they gain "highly specialized industry knowledge," placing themselves "between critics and fans, combining the acumen of the former with the enthusiasm of the latter."[38] On the other hand, insiders gained this knowledge via extensive consumption constructed by, and on behalf of, Hollywood discourses. Special editions and bonus features amplified the DVD's cachet, where, according to McDonald, "the film is one part of a package allowing the buyer to get immersed in the film's historical and cultural legacy."[39] As newer digital distribution channels emerged, the DVD lost luster as a primary site for paratextual repurposing and driver of conglomerate revenue. The rise of digital video marketplaces popularized digital copies and rentals. Cable and satellite set-top boxes also encouraged on-demand rentals. And Netflix ushered in an era of streaming video libraries. 2016 was the first year in which digital and streaming revenue surpassed that of the DVD and its high-definition successor, Blu-ray.[40] Spending on home entertainment reached new heights in 2016 thanks to growth in every category *except* for physical releases.[41] With the fading interest in physical media, special features have been gradually removed from DVDs and Blu-rays, particularly versions available via popular rental kiosk Redbox, while streaming or digital releases usually lack them altogether.[42]

But the decline in physical discs has not led to an equal dearth in repurposing. In fact, the circulation of paratexts has only increased, with companies filling websites and platforms with a deluge of videos, photos, webisodes, producer blogs, games, podcasts, and branded content. As John Tomlinson argues, corporations have exploited the culture of "rapid delivery" and "ubiquitous availability" by promoting the idea that *more content*—particularly short-form mobile content—is better.[43] The digital turn has permitted networks to, as Joshua Green writes, "dis-embed content from the broadcast schedule and re-embed it within different contexts."[44] Max Dawson calls the distribution of webisodes and digital shorts the "unbundling" of television across screens and platforms.[45] Gillian Doyle similarly refers to this multi-platform migration as a "360-degree approach," including repurposing and the creation of content for specific distribution channels.[46] Multi-platform dispersal enables the natural circulation of paratexts among fans, which inspires further consumption of episodes and seasons, the primary texts.

One notable form of paratextual dispersal is transmedia storytelling, which, as Henry Jenkins imagines, sees "integrating multiple texts to create a narrative so large that it cannot be contained within a single medium."[47] Unlike the unbundling of webisodes or diffusion of publicity interviews, transmedia storytelling compels audiences to engage with each piece of the storyworld to grasp the narrative. Jenkins locates transmedia as an authentic cross-platform experience where content is unified by a creator's vision. This approach contrasts with multi-platform strategies where people work separately to make content for different outlets and audiences. The web has deepened the appeal of transmedia storytelling for sprawling storyworlds. Episodes of *Heroes*, a serial about superpowered people, were appended with online comics and webisodes filling in backstories; websites detailing fictional organizations; character blogs and fan wikis; and episode commentaries.[48] *Lost*, meanwhile, was buoyed by an alternative reality game (ARG) known as *The Lost Experience*. Launched with clues embedded in commercials for Jeep and Sprite and expanding to websites, phone numbers, and videos, *The Lost Experience* revealed vital pieces of information about the DHARMA Initiative, a secretive organization dedicated to examining the island and the mystical numbers.[49]

Although the logic of transmedia storytelling aligns with television's shift to what Jeffrey Sconce calls "the crafting and maintaining [of] ever more complex narrative universes," it also promotes an idealistic vision of collaborative industrial practice.[50] Indeed, despite the involvement of the respective showrunners, *Heroes* and *Lost* exhibited the challenges of transmedia work. The *Heroes* team won a Creative Arts Emmy for its immersive online experience, but all of the-critical story material was shared during televised episodes, positioning the transmedia as nonessential content targeted at only the most attentive fans. Similarly, though *The Lost Experience* answered long-running questions about the DHARMA Initiative, those answers were never addressed in the main series. Instead, episodes contradicted transmedia reveals altogether.[51] Even when producers have good intentions, transmedia stories often function similarly to tie-in novels, comic book extensions, and video games: to affirm the importance of the primary text. In network television, this affirmation serves to sustain interest *between* episodes.[52] Moreover, as multi-platform and transmedia tactics connect narrative dots, they also shrink the range of potential viewer interpretations. Dawson contends that networks used "video abridgements" to construct a "paratextual scaffolding" around increasingly intricate and sprawling serials to "reinforce, redact, valorize, and demystify" the narrative.[53] Whereas behind-the-scenes DVD features explain *how* a film or series is made, more modern repurposed material explains the *what or why* of a narrative against which viewers must negotiate.

With repurposing in mind, we can draw a line from DVDs to multi-platform and transmedia storytelling to two-screen experiences. Story Sync took core traits of special features (behind-the-scenes tidbits, production diaries, commentaries) and multi-platform or transmedia strategies (narrative extensions and clarifications) and, using Green's term, "re-embedded" them into the "flow logics" of conventional live television.[54] Kim Bjarkman asserts that television has always been "a medium that is at once pervasive and scarce," but the rise of DVDs, DVRs, and streaming video have made it easier for viewers to stave off this ephemeral nature and retain all the programming they want.[55] Story Sync inverted this pervasive/ephemeral tension, deploying paratextual material to reconstitute meaningful ephemerality. This content repurposing promised fans the chance to take a more active, participatory role in unlocking new information or solving narrative puzzles. But for Hollywood, the goal is the same: to charm engaged fans with new information that sustains interest in main texts and revenue streams.

RELIVE KILLER MOMENTS:
THE ORIGINS OF STORY SYNC

By 2012, the enthusiasm for second screens and Social TV had reached a new zenith. As networks formalized a live-tweeting strategy, their corporate parents invested in other Social TV products. Viacom, Time Warner, and Comcast partnered with Zeebox, a UK start-up that mixed video clips, real-time chatter, an algorithmic programming guide, and sponsored integrations.[56] In 2012, Yahoo! purchased IntoNow, which synched to television audio to offer bonus information about the on-screen content, to improve its software for smart TVs.[57] Time Warner, CBS, and Viacom also partnered with companies that allowed viewers to "check in" to series in exchange for rewards. The similarities between these Social TV products did not matter to media conglomerates. As HBO's senior vice president of digital platforms, Alison Moore, said, these partnerships functioned most distinctly as another way to "promote its content wherever its viewers are."[58]

Social TV trials extended to branded second-screen experiences built around single series or networks. In 2010, USA Network introduced an app for comedic procedural *Psych*, including extended episodes, promotional videos, space for real-time conversation, and social platform integration.[59] Fox and producers of the competition series *The X Factor* introduced *The X Factor Digital Experience*, a live second-screen product with interviews, live questions submitted via Twitter, and backstage access.[60] Live sports coverage also inspired multi-screen expansion. Major League Baseball's At Bat, a partnership between its digital and television divisions, quickly grew into a success among fans.[61] NBC introduced two apps for the 2012 London Olympics, one for event livestreams and another for pertinent stats and information.[62]

Before 2012, AMC had sporadically produced digital extras for its acclaimed dramas *Mad Men* and *Breaking Bad*. But the immense popularity of *Walking Dead* catalyzed AMC's interest in multi-screen content that would support the zombie gore and narrative twists and keep people watching live. The first attempt was "watch-and-chats" on AMC's forums. There, the conversation was facilitated by moderators who also provided bonus material to the most active participants. AMC executive McKean praised the reach of this synthesis of chatter and digital extras, using an example of a *Walking Dead* watch-and-chat poll result that spread into fan circles via Twitter.[63] The success of watch-and-chats inspired Story Sync, a product aimed at triangulating conversation, bonus content, and live television. Announced before the second half of *Walking Dead*'s second season, Story Sync was situated as a live, fan-first experience:

> Attention *The Walking Dead* fans: Looking to immerse yourself deeper into your favorite TV series? . . . *The Walking Dead* Story Sync, a live, interactive experience that allows you to vote in snap polls, answer cool trivia questions, and re-live tense killer moments via video clips during the premiere broadcast of the latest episode. Think Shane should have stayed away from the barn? Vote in Story Sync's

Judgment Poll. See a particularly gruesome walker kill? Rank it on the Gore Gauge. As always, a live Watch & Chat lets you join fellow fans in the running commentary on that night's show.

The press release announcing Story Sync addressed viewers as engaged, social fans, with fandom pared down to a few core activities: voting in polls, answering "cool" trivia questions, ranking brutal zombie kills, chatting with fellow enthusiasts, and rewatching scenes. Story Sync was framed as "immersive" and "interactive," allowing fans to go "deeper," but on AMC's terms. The release also signaled Story Sync's focus on the live viewing experience. It twice mentioned live and alluded to liveness through phrasing like "the latest episode," "during the premiere [or first] broadcast," and "that night's show." This framing conveyed that the best way to use Story Sync—and the best way to watch *Walking Dead*—was live when the episode first aired.

While the first publicity cycle for Story Sync centered on fan-friendly buzzwords, AMC shifted strategy when building the product for *Breaking Bad* later in 2012. Story Sync retained polls, trivia, and interactive minutia but also integrated more "Photo Flashbacks" and "Thematic Callbacks" to prior episodes, dangling plot threads and pop culture references.[64] The press release echoed the shift with details on Story Sync's newer features: "Disapprove of Walt poisoning Brock? Vote in Story Sync's 'Judgment' poll. Think the bomb strapped to Tio's wheelchair is ingenious? Rank it on the 'Mastermind Meter.'"[65] These features previewed Story Sync's reiteration of the complicated morality of the *Breaking Bad* storyworld. The series pushed viewers to consistently re-evaluate their opinion of Walter White (Bryan Cranston), a teacher-turned-drug kingpin who began making methamphetamine to cover medical expenses but schemed his way into a larger operation. While showrunner Vince Gilligan was insistent about the moral decay of his central character, White became a seminal figure in television's "anti-hero" era, with many rooting for him to succeed. The sample questions implied that viewers would "disapprove" of the poisoning of young child Brock but find the strapping of a bomb to a wheelchair "ingenious." That one of the features was called the "Mastermind Meter" stressed that viewers would be, if not outright sympathetic to White, at least compelled by his choice to "break bad." With the Judgment Poll and Mastermind Meter situated as selling points, AMC prioritized an approved form of engagement with the text over engagement with other fans.

Adapting Story Sync to *Breaking Bad* led to new media coverage. In a 2012 CNN report, McKean revealed AMC's method to ensure the most robust Story Sync content: partnering with the writers. "We collaborate closely with the writers' rooms and production teams of the shows for every episode. In fact, the *Breaking Bad* one, we had someone from the writers' room actually scripting it," he said.[66] By identifying the contribution of "someone" from the writing staff—likely an assistant or script supervisor—McKean tried to instill Story Sync with deeper meaning. While Jenkins advocates for collaborative authorship to create a transmedia story, Elizabeth Evans argues that detecting authorship is more complicated when a franchise is spread across platforms.[67] Evans explores the idea of "institutional" authorship where broadcasters employ their vast resources to craft a transmedia project.[68] I would contend that institutional authorship also applies to writers' rooms, particularly regarding a secondary product like Story Sync. McKean tried to assure fans that Story Sync was not just a marketing tool with a *Breaking Bad* license; instead, it was an official paratext affiliated with the thriller's vision and voice.

To this end, McKean affirmed that, despite the broader discourse about multitasked viewing, Story Sync intended to deepen, not distract, from the television experience: "We're pretty judicious.... We literally watch the show like, five times, and we're constantly adjusting it. We're not asking you to click a button in the middle of someone getting shot, or a complex conversation that you should be paying attention to.... The second screen is connecting some dots that people might not connect for themselves. Our goal is to enhance the experience without making a new story."[69] While not precisely the "movie magic" that Klinger writes about the DVD feature, these comments suggested that the meticulous approach producers take to crafting episodes also translated to two-screen content.[70] McKean accentuated the utility of Story Sync without marginalizing the experience of watching new episodes. He stressed that AMC would not ask participants to disrupt their viewing of scenes to which they should be paying attention. On the contrary, Story Sync would make connections that viewers might not recognize on their own. McKean promised that the depth of AMC programs dictated bonus material to connect narrative or thematic dots—but not so much that multitaskers would be even more distracted. Again, this approach aimed to push viewers toward fidelity to the events on the primary screen.

McKean's comments embodied the paradoxes of Story Sync. On the one hand, *Breaking Bad* and *Walking Dead* were so complex that AMC's digital production team had to work aggressively *not* to distract viewers from the television screen. On the other hand, McKean suggested that Story Sync was *needed* because it connected dots for those who could not keep track of the litany of references within an episode. As Caldwell argues, digital technology has made repurposed content an "efficient" way to "add value" to existing properties.[71] Media companies must strike a delicate balance with this repurposed content. It must add enough value to attract consumers but also remain supplemental enough not to disturb more lucrative revenue streams. Across the early integrations into *Walking Dead* and *Breaking Bad*, AMC presented two distinct versions of Story Sync: one driven by fan-oriented interactivity and the other crafted by writers to complement the complex series playing out on the primary screen.

REPURPOSED ORIGINS: BEHIND THE SCENES AND ACROSS THE TEXTS WITH STORY SYNC

The split functionality of Story Sync was visible in its core features. While Story Sync initially expanded to a few seasons of other AMC dramas, including *The Killing* and *Turn: Washington's Spies*, it mostly remained linked to the *Walking Dead* and *Breaking Bad* franchises, including expanding to spin-offs *Fear the Walking Dead* and *Better Call Saul*. Meanwhile, other programs like *Hell on Wheels*, a historical epic about the US railroad system, or the crime drama *Low Winter Sun* were never given the Story Sync treatment. With that said, the layout of Story Sync did not change much between 2012 and 2017. The content—a photo or video overlaid by additional text or clickable prompts—appeared in the top two-thirds of the screen. The bottom third was made up of an informational preview bar, mapping future material. Story Sync content can be divided into four categories: 1) production and behind-the-scenes tidbits, 2) intertextual references and franchise ties, 3) interactive questions and polls, and 4) ads and product placement.

Story Sync's behind-the-scenes tidbits, dispensed in images and videos, most emulated the DVD special feature. Given that AMC aimed to keep attention on the television screen, the static images intended to offer quick bursts of non-distracting information. The Story Sync for *Walking Dead*'s season six finale, "Last Day on Earth," presented an inside look at Rick Grimes's (Andrew Lincoln) rifle as it appeared on television. Introduced with the bluntly described "Weapon" on the progress bar, the specs of Rick's rifle were broken down into simple descriptors: cartridge size, action, rate of fire, barrel length, weight, and sights. The information was paired with a close-up shot of the weapon, pushing viewers to compare between the static version and the "live" version in Rick's hands. The rifle revealed neither Hollywood magic nor production insight, but in presenting it as "immersive" information, Story Sync infused it with importance. As Caldwell claims, DVD features contribute to "the illusion of meaningful informatics," where producers frame any behind-the-scenes material as notable *by merely presenting it* to viewers, with little "meaningful outcome" in return.[72] The weapon linked the series and the real world, offered a token to those invested in verisimilitude, and played on the allure of personal survival in a zombie apocalypse. It, therefore, reiterated the value of weapons to the *Walking Dead* storyworld.

The videos, embedded into the flow of Story Sync during breaks, included talking head-style interviews with writers, producers, and stars detailing production practices, rationalizing story choices, and explaining the experience of working on a successful series. The "Last Day on Earth" Story Sync showed prominent figures walking the audience through both the final episode and the season as a whole. "Wrapping Up Season 6" described key plot twists in generalities but more pointedly detailed the emotional toll of the events on the characters. The clip began with dramatic and somber music before star Lauren Cohan, nearly in tears, said, "I don't know how it can be any worse than this." Fellow actors spoke in platitudes about the narrative stakes, noting, "Not everyone is fit for this world," "This is the first time that there's no way out," and "The game has changed." Ostensibly a plot summary, the segment tried to provoke an emotional reaction from viewers without spoiling details. The tone positioned *Walking Dead* as a series of exceedingly intense events, to the point where actors were emotionally moved when explaining the experience of making it. This tactic repurposed a Hollywood trope of the crew speaking about the hardships of filmmaking, but it also emphasized the program's survivalist themes and emphasis on character deaths. Klinger writes that DVD features have an "instant built-in and changeable intertextual surround that enter into [a text's] meaning and significance for viewers."[73] Parker and Parker similarly refer to features producing "a more self-conscious attitude" in viewers.[74] This intertextual surround and more self-conscious attitude can point toward preferred readings, particularly when bolstered by dramatically delivered talking points from industry representatives. By framing *Walking Dead* (both the story and the production) as emotionally and physically grueling, the cast and crew melded those versions of the series to sell a specific and compelling creative vision.

The second video, "Inside Episode 616," equally relied on emotional appeals and thematic reiteration but also moved further into the explanatory territory. Rather than taking viewers inside the production of the episode, the clip featured cast and crew members explaining key themes and character mindsets during tumultuous plot events. Executive producer Greg Nicotero spoke over clips of infamous comic villain Negan, the character whose anticipated introduction came at the end of season six: "Negan knows he's in control. Negan knows everything that he needs to know." Fellow producer

Gale Anne Hurd followed with more context: "Negan's retribution is swift, it is lethal." Finally, producer Denise Huth voiced the audience's assumed reaction: "Even before they go running out into that clearing . . . the audience feels, 'Oh no. Go back. *Go back.*'" Though products like Story Sync are often pitched as catalysts for participatory fandom, here the focus was not on active engagement with the storyworld. Nor was it on production revelations as one might see on a DVD set. Instead, the knowledge shared with viewers focused on interpreting the series a certain way, aligned with the producers' vision. Brookey and Westerfelhaus claim that the self-contained DVD package collapsed the boundaries between a film and its promotional material into a single "*intra*textual relationship."[75] Story Sync deepened this intratextual relationship as the information was synchronized directly into the live experience and left viewers far less time to process their interpretations of plot events or character motivations. The somber yet conversational tone of the clips was strategic. Any levity would undercut the subdued tone of the series and disrupt the flow of the thematic reiteration.

In fact, while AMC released more conventional behind-the-scenes videos outside on its website, the two-screen presentation focused far more on intertextual surround and franchise links. For *Walking Dead*, a program based on a long-running comic, the Story Sync interface was a useful platform to visualize connections between television adaptation and source material. Relevant comic panels emerged on the second screen as the characters appeared in the episode. These comic inserts, titled "Graphic Origins," gave non-comic readers the vital information needed to understand the source material or primed comic readers that a recognizable event was about to happen on television. The panels also referenced the specific issue where the character originally debuted. The Graphic Origins for "Last Day on Earth" showed that Negan arrived in comic issue #100, as well as that the character's first lines of dialogue in the comic closely mirrored his opening lines on television. These moments signaled to viewers that particular television sequences were important merely because of their existence within the source material. They also assured comic readers that the creators of the television series planned to honor the source text. While all transmedia or multi-platform stories face the challenge of extending canon to new spaces, adaptations are particularly fraught because fans have higher expectations about characters or plot points. These connections made significant nods to the larger storyworld and further underscored Story Sync's explanatory power.

Another Graphic Origins went beyond character introduction. During a scene where Cohan's Maggie appeased a distraught Rick that she still believed in him, Story Sync displayed a similar scene from the comic, with

Maggie declaring, "... I believe in Rick Grimes." For the most knowledgeable comic readers, the visual link between the two statements noted the significance of an anticipated moment of adaptation. For nonreaders, Story Sync delivered no additional narrative context for the panel, including any description or confirmation that the woman in the illustration was indeed Maggie. These examples recall Jason Mittell's "orienting paratexts," which help viewers better understand a storyworld "from a distance," but Story Sync exemplified how they can be used in real time to contextualize or explain on the fly.[76] More important, they show the importance of synchronization between two screens. Story Sync helped nonreaders learn more about the comic book roots of the scene, thus infusing it, and the primary screen, with more meaning.

Story Sync also delivered orienting paratexts through "Flashback," which displayed past moments deemed pertinent to the action happening in the current episode. When a scene in "Last Day on Earth" mentioned a horse, Story Sync showed an iconic shot from the pilot episode where Rick attempts to escape a zombie horde on horseback. The text description of the pilot sequence was quite direct: "Rick rides into downtown Atlanta on horseback. (Season 1, Episode 1, 'Days Gone Bye')." Flashbacks were not interactive—viewers could not click into them to watch an expanded clip of the sequence—and they, like Graphic Origins, remained on-screen briefly. Flashbacks thus did not just connect narrative dots. They also celebrated seminal moments in franchise history, with or without narrative relevance. Sconce argues that, for modern serials, the process of world-building is as vital as episodic storytelling.[77] But on Story Sync, content turned inward toward intra-franchise references rather than outward to construct a larger storyworld. Another Flashback, labeled "Before & After," offered a line of dialogue from Abraham (Michael Cudlitz) that harkened back to a prior episode. To make the link explicit, Story Sync presented a split-screen image of Abraham and an awkwardly positioned chunk of text: "'I see rain comin', I'm wearin' galoshes.'—Abraham to Glenn on starting a family (Season 6, Episode 11, 'Knots Untie'); 'I could. Now. Just so you know.'—Abraham, to Sasha (Season 6, Episode 16, 'Last Day on Earth')." Compared to the Graphic Origins or Rick and the horse, this image worked harder to link two moments. The second screen explained character development not fully realized in the episode. Alert viewers may have noticed the connection between the scenes, but AMC did the interpretive work for them in real time.

The Flashbacks and Before & After segments for *Better Call Saul* naturally addressed the its status as a *Breaking Bad* prequel. However, because *Saul* follows a prequel story with a distinctive tone and genre in the legal serial,

Story Sync also often accentuated its links to US law. The Story Sync for the season two finale, "Klick," included sections called "Legalese" and "Criminometer." The first offered definitions for legal concepts like the Hippocratic oath, while the second celebrated the episode's "most pivotal crime committed," complete with its statute code and punishment (e.g., imprisonment). Similarly, early episodes of *Turn* offered digital maps charting the progress of its characters in real historical events along with excerpted analysis of the events from textbooks. Here, Story Sync situated the fictional series within the "real world," or at least a consumable form of it. The source material was no longer from a comic panel or fictional locale but instead legitimate laws and sites of war, thus performing devotion to verisimilitude. This is a standard Hollywood approach, where producers, stars, or studios express respect for source material or historical accuracy. Whether affirming knowledge that experts already have, or providing new contextual information for the nonexperts, these moments positioned Story Sync and AMC series as dedicated to authenticity.

These features posited that AMC storyworlds were complex enough that they must do what Will Brooker calls "overflow" to secondary screens.[78] Brooker examined early multi-platform ephemera in *Dawson's Creek* websites that expanded its storyworld between episodes, offering updates on the lives of the fictional characters and the location of Capeside. Though *Creek* overflow content expanded the storyworld, Story Sync's synchronized content was more self-referential, linking back to minutia in established storyworlds. In this way, despite storyworlds overflowing onto many screens, Story Sync reduced its intertextual references into trivia. Years of source material, adaptations, and real history were transformed into still images and short sentences. Klinger argues that DVD features present viewers with repeatable, easy-to-recall talking points that conflate expertise with "the accumulation and dissemination of the smallest details involved in the production of media." Trivia has long been "a source of popular expertise" for fans, enabling the media industries to "dutifully produc[e] massive amounts of this kind of information."[79] Neither trivia nor fans' desire to search it out are negative indicators of industry-fan relationships. Not only did Story Sync's trivia involve rudimentary recognition of intertextuality, but it also sat alongside other material that pushed a specific narrative interpretation. Writing about the experience of watching television on a computer, Brooker claims that episodes are "in competition with various other equally-demanding 'screens' within a larger screen."[80] This experience facilitates entry into multiple windows for research, where viewers move to a web browser to search for an actor's filmography or production insights. AMC promoted Story Sync as a solution to this multi-screen navigation so viewers could participate in a streamlined experience between screens. The intended byproduct of the tactic, however, was that committed participants would not leave Story Sync at all.

Story Sync tried to be many things at once. It aimed to display the depth of AMC programming but only in small segments. It offered talent interviews, not for insider production tidbits but rather to shape approved episode readings. And while it aimed to inspire fan drilling into beloved serials, it also wanted to contain fans within the live two-screen enclosure and guide them toward callbacks, intertextual references, and trivia.

GORE GAUGES AND CELEBRITY CAMEOS: STORY SYNC'S INTERACTIVITY

Given that Story Sync was inspired by fan discussion on AMC's website, the product predictably featured a live chat room known as "Fan Reactions." But Fan Reactions could not be accessed at the same time as other Story Sync content, meaning that viewers had to choose between conversation and preprogrammed interactive features. Beyond the chat room, Story Sync encouraged viewers to take part in quick polls that reinforced the programs' most relevant touchpoints. Interactive components for the *Walking Dead* franchise included the "Tactical & Morality Matrix" and the "Survival Matrix," which asked participants to judge the arduous choices of the characters. These matrices were visualized by a color-coded four-quadrant grid and punctuated by questions relevant to the television scene. The Story Sync for "Last Day on Earth" included a Tactical & Morality Matrix prompting viewers to decide if, "Letting the Saviors' [a villainous group] victim die . . ." was "Tactically Right and Moral," Tactically Wrong but Moral," Tactically Right but Immoral," or "Tactically Wrong and Immoral." *Fear the Walking Dead*'s "Date of Death" episode likewise prompted viewers to evaluate a character's request to be mercy killed. The choice was between "Civilized and Justifiable," "Primal but Justifiable," Civilized but Unjustifiable," and "Primal and Unjustifiable." The matrices appeared briefly on the second screen to force viewers to make a quick gut decision about difficult, violent, or deadly acts. Although viewers were able to see how everyone voted, Story Sync gave no space for reflection or debate about the supposedly challenging choices. It instead traded on appreciation for characters and the storyworld to reiterate germane themes. Viewers surely developed a connection to the characters whose actions they judged or imagined how they would personally operate

Tactical & Morality Matrix from a *Walking Dead* Story Sync asking participants to vote on characters' choice to let an antagonist die.

during a zombie apocalypse. They thus evaluated how characters fit inside both the framework of the diegetic universe and their own morality.

Story Sync distilled this focus on morality in a feature called "Judgment." Viewers were prompted to pick between just two options—one positive and one negative—in appraising character decisions. For "Last Man on Earth," Story Sync asked viewers if they "believe[d]" Rick in a conversation with Maggie, or if he was "just comforting" her. The choices were color-coded, with the former in green and the latter in red. Obviously, this implied that one option was morally right and the other morally wrong. This tension is vital to *Walking Dead*'s themes, as Rick, a classic hero type, consistently wrestled with how to protect his group, even if it meant lying or sacrificing few to save many. Story Sync externalized Rick's internal debate, challenging people to evaluate his choices in real time. Despite the dichotomy of the evaluation, the Judgment feature upheld *Walking Dead* as a complicated moral universe, where decisions regarding how to handle interpersonal conflicts were just as tough as navigating violent attackers. The product's baseline interactive features bolstered Story Sync's synchronized reiteration. Throughout the episode, Story Sync tracked viewer responses to the Morality Matrices and Judgment questions, and, in the end, generated a match to a character whose morality the viewer most mirrored. As Story Sync asked participants early on, "Who are YOU most like? Will you SURVIVE this episode? Answer questions within this SYNC to find out." Caldwell argues that DVD features "embed a rich mix of critical analysis and on-screen theorizing" to "'negotiate' critical reception in the culture at large."[81] By personalizing the call to action, Story Sync pushed participants to identify with the tough choices of the characters and the moral complexity of the storyworld *even more* than it asked them to recognize the expertise or depth of the world. This reached participants before they consumed outside interpretations (from critics or other viewers on social media), thereby solidifying an approved authorial vision for a given sequence or episode.

Another notable feature of *Walking Dead*'s Story Sync was the "Gore Gauge," which asked viewers to rate the degree of violence inflicted onto a zombie. The Gore Gauge for "Last Day on Earth" showed one of the "walkers" trapped in chains that wrapped both around and through its body. As a high-definition close-up on the emaciated, bloody walker filled the screen, viewers were pressed to pick between "Barely Bloody," "Some Splatter," "Guts Galore," "Major Carnage," and "Total Bloodbath." Gore Gauge represented a core reason why *Walking Dead* became so popular: people enjoy watching zombies being destroyed in a multitude of ways.

Together, these Story Sync features exemplified the distinction between interactivity and participation. Jenkins asserts that interactivity "refers to the ways that new technologies have been designed to be more responsive to consumer feedback," wherein all modes of interactivity are "prestructured by the designer."[82] Devices and platforms are made to be interactive, and users can gain a measure of control of their experiences. Participation, meanwhile, emerges from cultural and social practices and collaborations within everything from popular culture to politics.[83] Still, the media industries conflate interactivity and participation, with technology situated as the conduit for meaningful participation. This utopian viewpoint has long undergirded the discourse about technology's impact on the passive experience of television viewing. Jo T. Smith approaches the interactivity of DVD with equal skepticism, arguing that it "involve[s] a mode of address that invites us to *do* something with our media objects."[84] This ideology of *doing something* was central to the Story Sync experience. AMC entered into the Social TV ecosystem with reason to believe that viewers wanted to do something with their devices while watching. But while many Social TV products sold participation in the form of conversation, AMC pitched a structured but functionally interactive experience propelled by programming.

Part of Story Sync's emphasis on live viewing was to convince viewers to pay attention to the bevy of product integrations, branded segments, and cross-promotions. The promotional content included integrated sponsorship deals, publicity for other AMC products, and cross-promotion involving celebrities unaffiliated with the network. In the first category, Story Sync generated an ad to coincide with the commercial break on television. Sometimes Story Sync displayed conventional static banner ads and thirty-second video spots. Other times, the ads more explicitly synchronized to a production integration within the episode or the spot on the television screen, delivering a double-shot of product placement. Therefore, the synchronized reiteration strategy extended beyond a given program's themes to include a conventional commercial sales pitch. While DVD bonus materials seldom included sponsored messages, the promotional focus was mainly on Hollywood. Transmedia experiments like *The Lost Experience*, however, have been more regularly incorporated into the promotional mix, with pieces of the puzzle inserted into ads for global brands. The real-time functionality of Story Sync offered another way to attempt to overcome time- and place-shifting and deliver key demographics to sponsors.

AMC executive McKean was direct about the ad integration. He unambiguously celebrated Story Sync's potential role in buoying live viewership and attracting companies with ad budgets to spend: "[I]t's very popular with sponsors. . . . We're a business, so certainly we're trying to drive our business. But hopefully, we're trying to drive our business by creating great experiences for viewers and fans."[85] McKean spoke to all of AMC's audiences; in his mind, Story Sync offered "great experiences for viewers and fans," but he also admitted the benefits of a digital enclosure full of ads. For AMC, then, repurposing also applied to its distribution of ads and sponsored content. In relying on familiar strategies, Story Sync typified Green's point that most web television products "negotiate an identity as an evolution of broadcast television, rather than necessarily positioning [themselves] as an object that breaks from it."[86]

Story Sync also featured cross-promotions for pertinent franchise content and celebrity partners. *Fear the Walking Dead*'s "Date of Death" Story Sync included a promotional plug for its affiliated mobile game *Dead Run*. Similar to the ad format for outside products, the mobile game spot presented a static image with a faceless hero firing a gun at a walker. But unlike many of the ads on Story Sync, the image for *Dead Run* included a red box that contrasted with the black-and-white imagery, imploring viewers to "Download Now." Caldwell asserts that web users are "always one click away from going somewhere else and thus creating their own 'unruly' migration or 'flow' across the web." In contrast, the DVD experience "rewards consumer impulse and the possibilities of cultural gratification and distinction that are packaged in 'featurettes' and bonus tracks."[87] Generally, Story Sync combined these experiences to limit unruly migration to outside content, browser tabs, or devices—except in the case of ads for affiliate products like *Dead Run* where movement was not just allowed but encouraged.

Meanwhile, AMC used Story Sync to repurpose its cross-promotional deals with celebrities during its popular aftershows that break down shocking twists or big deaths.[88] When a celebrity was slated to appear on an aftershow, AMC sporadically asked them to participate "live" on Story Sync. The Story Sync for *Fear the Walking Dead*'s "Date of Death" began with a notice that one of the stars of HGTV's *Property Brothers*, Drew Scott, would be watching as well. Labeled "Watch Together," the announcement encouraged viewers to take part in Morality Matrices and Judgments to see how Scott answered the questions, indicated by the abrupt appearance of his face. Here, Story Sync navigated several levels of promotion: it stressed the value of synchronized live viewing and the intrigue of the morality-driven polls but also cross-promoted *Property Brothers* and plugged Scott's forthcoming appearance on *Talking Dead*. Story Sync instructed viewers how to hide Scott's activity, but it would still reappear in crucial moments. Any ads that AMC could push to attentive eyes were, as McKean said, only a bonus.

According to AMC, the audience for Story Sync was just "a tiny fraction of viewers, but they are the really committed ones."[89] Whereas many Social TV products tried to obscure their interest in connecting sponsors with consumers, Story Sync used patented strategies like product integration and

Fear The Walking Dead Story Sync promoting Drew Scott's real-time participation indicated by an icon featuring his face.

celebrity cameos to make ads and sponsored content, which were embedded into conventional television flows, more appealing. Still, the presence of certain sponsored elements like ads for franchise mobile games or celebrity participation were also pitched as an explicit benefit to the fan-forward Story Sync experience.

"AMC JUST TROLLING US AT THIS POINT": FAN RESPONSE TO STORY SYNC

Even before its termination by AMC, the ephemerality of Story Sync made it challenging to gauge viewer response in real time. The chat room was separate from the main content and inaccessible after live episodes. Though this potentially generated more conversation during the multi-screen experience, it creates challenges in analyzing participant feedback. To contextualize how people engaged with and discussed Story Sync, I offer commentary from users of Reddit, the popular news aggregation and discussion board. Reddit has become a space for the narrative sleuthing encouraged by Story Sync, and "redditor" investigations have made it on Hollywood's radar. *Better Call Saul* producer Vince Gilligan admitted that users cracked a hidden message in the titles of season two episodes long before he intended to reveal it.[90] Series like *Big Little Lies* and *Westworld* also have had plot twists solved by eagle-eyed redditors.[91] There are obvious qualifiers to this commentary. Reddit posts did not respond to the specific Story Sync material noted in prior sections, and while the observations contextualize Story Sync, those who both used the product and talked about it on Reddit represent a small, active segment of the fanbase. Reddit comments should thus be understood as one thread of response to Story Sync.

Given redditors' tendency to try to solve fictional narratives, most of the conversations about Story Sync focused on its informational value. Redditors on specific "subreddits" for *Walking Dead*, *Fear the Walking Dead*, and *Better Call Saul* posted remarks about just-concluded Story Syncs, including screenshots from their devices. These threads were given the "[SPOILER]" tag, implying that evidence from Story Sync would give away relevant plot details. The threads were introduced with titles like: "AMC StorySync giving us a hint?"; "Thank you Story Sync for a good find"; and "A great catch you might have missed if you don't use Story Sync." Sharing material that filled in narrative blanks, nodded to past plot developments, or presented amusing intertextual references, fans underlined Story Sync's function as a critical tool in the hunt for vital narrative connections. The "great catch" thread highlighted a *Better Call Saul* Story Sync callback to *Breaking Bad*, with some believing that the presence of a *Breaking Bad* scene on Story Sync confirmed an imminent development on *Saul*:

> **madhjsp:** Aw snap, so the Salamancas are gonna come after Mike. Maybe that's the beginning of how he comes to meet Gus.
> **kaztrator:** New theory: Gus has Mike drown Hector to the point of brain damage.
> **Audihoe:** This is what I'm thinking. Or maybe hector catches a bullet but it just grazes him.
> **mattyn33:** This is a great connect. At least we know Mike will be ok. The closer Mike gets to the cartel … the closer he gets to meeting Gus.
> **Brandeis:** Good catch, but there were no witnesses left behind by Mike. A witness has to *see* or *hear* something or someone. By definition, the truck driver isn't a witness.[92]

In this instance, Story Sync material helped confirm or deny ongoing theories that redditors had about *Saul*'s prequel narrative, especially its connection to "future" events that already occurred in *Breaking Bad*. Mittell's drillable metaphor implies that fans "mine" for information that has been strategically embedded in the narrative.[93] *Saul*'s status as a prequel inevitably dictates that fans would dig into it to find connections to *Breaking Bad*. Story Sync delivered additional evidence to deepen the digging.

Story Sync material also gave redditors something to argue against. A post from the *Breaking Bad* subreddit presented a screenshot from Story Sync asking viewers to speculate about how Walter White's story might end provoked debate among redditors offering their interpretation based on the evidence presented from prior episodes:

> **groganjosh:** I think he will definitely be rich, in the flashforward he tips the waitress $100. No poor man does that.
> **hamza780:** He could be dying (or close to death) and just gives away any remaining money he has.
> **MyPhantomile:** With no family and fallen empire, what use is money to a dying man? I am of course making assumptions, but it's seeming to be a likely scenario.[94]

Here, Story Sync's Judgment incited a debate that undoubtedly occurred on the *Breaking Bad* subreddit countless times before the series ended. The paratextual material served as additional context to that ongoing conversation, with fans engaging in the type of discussion about morality promoted by Story Sync. But not all redditors were satisfied with how Story Sync prompted such a structured question in evaluating White's potential fate:

> **elbruce:** I guarantee you the person who set up this poll didn't know the outcome and therefore didn't know all of the possibilities on the table.
> **CapatchaInTheRye:** Well, those are more or less the only 4 possible permutations of those 2 things. You don't really need to know what's going to happen to put together a table like this.
> **elbruce:** Yes, I'm aware of that. The poll is pointless as it's [sic] set up.

For these users, Story Sync distilled an important narrative question and moral concern to a simplistic choice. These fans did not believe that Story Sync delivered meaningful engagement and instead settled as a "pointless" resource for those looking for a worthwhile debate. The idea that AMC used Story Sync to present inessential information popped up on Reddit a handful of times, particularly in discussions about whether or not it spoiled upcoming moments or deaths.

In another example, a *Walking Dead* fan shared a screenshot that their answers in Morality Matrices and Judgments resulted in Story Sync identifying them as "most like" Abraham and that they did not survive the episode. Coincidentally, the end of the episode implied that Abraham was in danger. The user believed that this confluence of events proved Abraham's death. A debate ensued about the mechanics of Story Sync's polling, with users accusing the original poster of editing the screenshot to convince the others that Abraham had indeed died. Eventually, redditors critiqued the mechanics of Story Sync, with one user noting, "AMC just trolling us at this point. Smh," and another replying, "They have been for a while."[95] The allegations were not only that Story Sync spoiled a significant cliffhanger but also that it actively toyed with expectations about potential deaths. To that end, Story Sync disrupted fans' attempts to drill into *Walking Dead* and obscured a mystery they were trying to solve.

Other examples debated Story Sync's connection to source material and author figures. On the one hand, *Walking Dead* fans celebrated Story Sync's shift in language on the "Threat Level Meter" from "Severe" to "Effin Severe" in honor of the first appearance of Negan, a character known for his use of explicit language.[96] The thread showed that redditors enjoyed that Story Sync changed the wording to affirm its fidelity to the source material. On the other hand, fans pushed back against Story Sync when it challenged franchise creator Robert Kirkman. A *Fear the Walking Dead* viewer shared a screenshot from Story Sync featuring fictional World Health Organization documents revealing explanations for a virus outbreak. This was an essential piece of evidence because Kirkman had sworn to never reveal this information. User Roninjinn asked about Story Sync's reliability in light of Kirkman's comments: "I could be wrong, but from everything I've seen, we have been told that Robert doesn't want to talk about the origin of the outbreak."[97] The ensuing exchange debated how previous episodes had failed to address this evidence, what assumptions viewers could make about the virus, and why the prequel seemingly planned not to answer an origin question.

> **ssjjfar:** It being a virus isn't something new though. They from the CDC already referred to it as one and that it was airborne. The origin he isn't going to talk about is where the virus came from/started.
> **Tario70:** I don't recall it ever being mentioned that it was airborne just that everyone was infected. . . . It's interesting that Kirkman doesn't want to explain it but that's what we're driving for & what many people want from *FTWD*.
> **Roninjinn:** Sadly too many people are expecting that explanation, which I agree isn't going to come, but we may get more pieces to the puzzle in little tidbits like this. I think it's fun to theorize. Talk about possibilities, and present/refute them in a civil manner. Just gets you involved more in the cannon [*sic*] verse, and characters IMO.

Users here spoke to Story Sync's explanatory power, even as they debated its legitimacy. They tackled questions of authorship motivated by both the spin-off and the multi-screen experience. The redditor introduced the topic with many qualifiers—"I could be wrong," "from everything I've seen," "we've been told that Robert doesn't . . ."—while respondents turned to the other evidence, found either in prior episodes or Kirkman commentary. Story Sync, on the contrary, offered "little tidbits" that made it "fun to theorize." These two examples showed that some fans appreciated when Story Sync underlined a program's faithfulness to source material but did not like information that conflicted with the franchise's auteur. If Story Sync affirmed their theories or admiration of the franchise, they would circulate it; if it did not, they would debate its validity.

Story Sync inspired a range of responses from redditors. Some gleefully presented its paratextual material as evidence in the broader pursuit to unlock mysteries, while others questioned its use in that pursuit. What is clear, however, is that fans of AMC programs did share and debate Story

Sync. Yet, the presence of Story Sync content on various subreddits also signaled that AMC's attempts to structure the live experience of fans—keeping them siloed within the confines of both screens—did not prevent fans from screenshotting, saving, and redistributing material for later. Grainge argues that television is "now less ephemeral in the evanescence of program content but much more ephemeral in the brevity of the promotional and paratextual forms that surround, mobilize, and give meaning to that content."[98] AMC embraced this by pitching the full Story Sync experience as ephemeral but surely expected that viewers would use digital tools to ensure that relevant content would live online as part of drilling down into the storyworlds. Nonetheless, this tactic situated Story Sync as an experience built around program and narrative material, not conversation or community. The chatter emerged elsewhere and still focused on how Story Sync matched existing assumptions about AMC programming.

CONCLUSION: OUT OF SYNC

Story Sync was yet another distribution channel for AMC to repurpose, disperse, cross-promote, and synergize. At times, Story Sync promoted the links to source material and flashbacks to prior episodes filled in narrative blanks, and viewers—hailed as fans—were able to complete an interactive experience that inspired further analysis and conversations elsewhere on the web. Mostly, however, the Story Sync features affirmed the depth and breadth of AMC programs, both narratively and thematically. References to source material or historical events positioned series as meticulously and authentically made, assuring viewers that producers had the franchise's best interests at heart. Meanwhile, brief videos and interactive components stressed vital themes and explicitly explained narrative developments. These components courted viewers as experts, promising an exclusive participatory experience. But more directly, these features aimed to structure participants' multi-screen experience and their reading of AMC programs. Altogether, the viewer experience fixated on information: the recognition of connections, the collection of trivia, and the proper interpretation of significant moments. The tactic went against predominant trends in the Social TV era, where many of the new products or practices were centered on real-time conversation and the sharing of multimedia content. Story Sync ultimately typified how the media industries tend to assess consumers. Even with emergent platforms for dialogue and the visible decline of physical media, the collectors of DVD box sets are now perceived to be collectors of information above all else.

The Social TV boom briefly inspired other networks to create second-screen experiences centered on programming. Showtime embraced the interactive directive with the familiarly named SHO Sync app in 2012 and later integrated its interactive content into smart TVs. Partnering with manufacturer LG, SHO Sync delivered "on-screen trivia, polls, and social features" directly into the episode, all enabled by a fingerprint-matching system.[99] HBO appropriated the DVD box set with "interactive features" for *True Blood*, *Boardwalk Empire*, and *Game of Thrones*. Available on streaming portal HBO Go, the features included cast and crew commentaries, interviews, explanations of particular references or source material, maps, and concept art.[100] This setup also allowed people to consume the bonus content with the episode on pause and then return to their viewing. HBO also created a social experience known as HBO Connect that added chat rooms and social feeds to HBO Go.[101]

Like with Story Sync, Showtime and HBO's multi-screen products scored immediate attention from the press. Nevertheless, despite the initial suggestion that these apps were part of the future, the response from viewers was less enthusiastic. A 2014 survey commissioned by the Consumer Electronics Association and National Association of Television Program Executives found that only 13 percent of people believed that synchronized second-screen content made the viewing experience "much more enjoyable." Though 67 percent of participants said that content improved viewing "somewhat," the survey revealed a lukewarm response to second-screen platforms.[102] By 2015, both Showtime and HBO abandoned these synchronous experiments. In explaining HBO's decision, executive Sabrina Caluori leaned on brand identity: "We are called 'Home Box Office' for a reason: we deliver cinematic-like experiences.... When you are in the movie theater, you don't use your phone. You are actually paying attention to the first screen. So why are we trying to distract you on the second screen?"[103] Showtime, meanwhile, utilized the "failure equals success" rhetoric of Silicon Valley with its cancellation statement. An anonymous executive said that SHO Sync helped the network "gain incredible learnings about what our fans were interested in within each show."[104] The comments are apparent attempts to spin failure as a learning experience, but even failures give media companies new data from which to build.

Indeed, HBO has continued to produce extensive bonus content for its tentpole dramas. Rather than embedding this content into one interactive package, HBO strategically uploads it to digital platforms, releasing scenes, explanations, and behind-the-scenes insights during the week between episodes. HBO has also developed a website that extensively details *Thrones*' textual universe.[105] As chapter 5 will describe, this is part of HBO's evolving digital footprint, drawing attention to programming beyond live episodes to position the network as a more socially authentic brand. Showtime too admitted that its Sync content would be better served "on platforms that already have a built-in audience," namely, Twitter, Facebook, and Tumblr.[106]

Similarly, AMC lost interest in Story Sync as part of its digital portfolio. Series that were tailor-made for the multi-screen experience, such as the alternative universe samurai action spectacle *Into the Badlands* and the comic book-sourced *Preacher*, were noticeably absent. By early 2017, AMC ceased production of Story Sync content without an announcement, leaving fans on Reddit to ask about the product's disappearance.[107] AMC executives have not been as transparent about their decision to conclude the Story Sync experience. Still, the network continues to produce new content for its websites and apps, including a virtual reality simulation for *Walking Dead*. The shift to something like VR indicates that AMC is always looking for new ways to engage its most dedicated fans through ersatz "interactive" experiences fixed mostly on its programming.

That remnants of these multi-screen products remain as part of new plans only underlines the media industries' commitment to repurposing. Story Sync shared a lineage with cultural products and industry approaches from prior generations that arrived with the same promises about interactivity and participation. While digital technology makes the processes of repurposing much easier, Hollywood still must experiment to find lucrative products for each new generation of consumers. Story Sync also exhibited that it is hard to unify disparate types of content when fans respect the voice of a singular author. It can be equally tough to convince fans that they must prioritize one platform over another in their preferred viewing experience.

Ultimately, Story Sync proved that television in the Social TV era became an enormous site of contention: between captive and multitasking viewership; between the live and the time- and place-shifted; between programming and conversation; between interactivity and participation; and between the ephemeral and the permanent. These tensions are not wholly conflicting poles. Instead, they have remained in industrial products and viewer experiences as all parties navigate the transition from broadcasting to on-demand culture. Story Sync tried to reconfigure television's ephemerality. On the one hand, the multi-screen product reconstituted the importance of single episodes, which are increasingly available across an array of platforms, by tethering them to fleeting ancillary content. But, on the other hand, Story Sync accepted that the bonus material could extend the life of episodes as fans saved it and used it to debate and excavate narratives in other digital realms. Repurposing is best understood as a way for the industry to combat these tensions because it stabilizes uneasy developments or unruly consumer practices. Every new platform is a chance to start again, and each new technological expansion offers another way to frame familiar tactics as disruptive.

Chapter 3

REWARDING VIEWING

Check-Ins and Social Productivity

Imagine watching the latest episode of your favorite television series, either live or on-demand. Now imagine that, instead of following typical Social TV protocol and posting your thoughts online, you were given a $20 gift card to a big box store. Or, at the end of a season's viewing, you were honored with the title of television Guru and mailed a dozen stickers featuring beloved characters. In a landscape of live-tweeting and bonus social content, these practices might seem unusual. But they were genuine, and briefly celebrated by start-ups, media conglomerates, and reporters alike. Gift cards, exclusive titles, and stickers were the centerpiece strategies for a collection of start-ups—GetGlue, Miso, and Viggle, most notably—that were known colloquially as check-ins. As the name suggests, the companies offered viewers rewards for watching television and reporting back to a digital platform using a device of their choosing.

Check-ins generally combined the features of two significant social media products: the communicative capability of Facebook and the location-based "self-reported positioning" utility of Foursquare, which enabled users to virtually check in at physical locations (restaurants, bars, and entertainment options) and score a litany of minor rewards.[1] Accessible via mobile devices and promoted by television networks, these products allowed users to search for entertainment options, check in, and report their tastes to fellow fans of a particular program or film.[2] GetGlue, Miso, and Viggle had slightly different check-in procedures, but each asked users to share their taste profiles, connect with peers, obtain points and rankings, and produce shareable content.[3] The range of activity and post-check-in prompts once again assumed that audiences wanted to go beyond the typical viewing experience—but within a tightly controlled environment. And, much like networks' live-tweeting campaigns and two-screen experiences, check-ins piggybacked on the ephemerality of live television by tempting users with maximum rewards and bonus points during the nightly primetime period. Checking in, then, was situated as yet another way to "do something" within the Social TV ecosystem and be rewarded in the process.

While the companies rarely released user information to the public, scattered reports suggest check-in participants fell within the key 18–49 age demographic.[4] Each platform rewarded user activity with prizes in different ways. On GetGlue, check-ins gave users an increasing number of points and digital stickers featuring series' promotional art, logos, or stars.[5] Additional check-ins and digital stickers allowed users to progress through a series of titles—Season Fan, Super Fan, Diehard Fan, Elite Fan, and Guru—and eventually receive hard-copy versions of accumulated stickers in the mail. Miso worked similarly, offering a running tally of points where users recorded one point for each check-in and a smaller digital sticker program. On Miso, more points gave users increased visibility throughout the platform. Viggle, meanwhile, disregarded stickers and instead distributed one point each minute of viewership tallied by "an anonymous digital fingerprint," which synchronized to sound waves emanating from nearby televisions.[6] Once users recorded enough points, they could cash them in for gift cards to retailers (Best Buy, Target) and restaurants (Subway, Starbucks), for consumer goods (headphones, tablets), or for tickets for Royal Caribbean cruise vacations.[7] All three products delivered points and rewards as part of the core experience, but bonuses came with strings attached—including watching specific series at specific times, watching unskippable ads, or playing sponsored mini-games.

The emergence of check-ins ran parallel to Twitter's rise as the critical Social TV space, as well as networks' attempts to develop their multi-screen apps. Reporters quickly positioned check-ins as the next big thing to alter the promotion and consumption of television. Indeed, of all the notable players in the Social TV landscape, check-in start-ups most represented the Silicon Valley ethos of utopianism and disruption as they angled themselves as leaner, more inventive alternatives to Facebook and Twitter. Nonetheless, in marking their territory against established social platforms, check-ins required funding from corporations across sectors. GetGlue raised more than $24 million from investors, including Time Warner, and collaborated with more than seventy-five networks and twenty-five studios. Miso aligned with DirecTV, AT&T, Yahoo, and Showtime, while Viggle partnered with consumer brands Hyundai, Lexus, Clorox, and McDonald's.[8] Along the way, Comcast, DirecTV, Yahoo, CBS Interactive, and Time Warner turned check-in investments inward, developing in-house platforms that eventually oversaturated the market for products that held niche appeal.[9]

Notwithstanding the glowing coverage from the press, check-ins never fully captured the attention of consumers. By late 2012, GetGlue's user base had grown to nearly four million users. Viggle had half as many. Miso, meanwhile, had flatlined with fewer than a million users and was purchased by Dijit, developer of another check-in product, NextGuide.[10] The hype gave way to volatility. A botched merger between GetGlue and Viggle in late 2012 paved the way for yet another Social TV start-up, i.TV, to swoop in, purchase, and rebrand GetGlue as tvtag later in 2013.[11] New ownership and a new name did not help; by the end of 2014, i.TV shuttered tvtag.[12] The products created by media companies similarly vanished with no fanfare. Viggle was the sole surviving product of the check-in bubble, with nearly ten million registered users as of 2017. Even so, it, too, was engulfed by Silicon Valley's predisposition for whirlwind acquisitions. Viggle purchased Dijit—the proprietor of NextGuide and buyer of Miso—in 2014, only to be bought by another reward-driven start-up, Perk, Inc., in 2016.[13] For a short time, check-ins were part of the industry's social promotional mix. Yet, despite their focus on unique rewards, check-ins never offered enough people a genuine reason to switch from more populated spaces.

This chapter examines the rise and unceremonious fall of check-ins. Drawing upon observations of publicly available user activity and my experience in 2012 and 2013, I describe the core features shared by GetGlue, Miso, and Viggle, including 1) the gamified point and reward systems, 2) a social experience focused on displays of fan knowledge and devotion, and 3) a prioritizing of live viewing and corporate partnerships. I position check-ins as a definitive Social TV product, one that imagined a type of user engagement between conventional notions of active fan production and simple consumption. Though the companies offered users activities to compete to procure points and rewards, and while some users worked collaboratively to game the check-in systems, these activities were not up to the level of what most would consider traditional fan production. Recalling John Fiske's visions for fan productivity—semiotic, enunciative, and textual—I offer a new category: *social productivity*, which blends real-time chatter and the creation of new, ephemeral material (brief reviews, embedded videos, and reaction GIFs) on social platforms. Like all cases of the Social TV era, GetGlue, Miso, and Viggle controlled this social productivity and yet promoted it as essential to the fan-oriented check-in ecosystem.

In analyzing check-ins' rewards systems, I reveal that they tried to mimic the conditions of the online "gift economy." Scholars have debated how the web's social dynamics could spawn a non-capitalist economy that thrives on camaraderie, collaboration, and gifts, including the free circulation of information.[14] The gift economy contrasts with the commodity or exchange economy. In an exchange economy, commodities shift to wherever profit can be found, and status is bestowed to those who have the most. Conversely, in a gift economy, gifts circulate via acts of generosity that intend to improve the community. Status, then, is afforded to those members who give the most.[15]

By ignoring this vision of how the internet could function, corporations regularly conflate the logic of gift and commodity economies. A notable example of corporate malfeasance of this type is 2007's FanLib. Armed with millions of dollars of venture capital funding and led by former Yahoo! executives, FanLib purported to give fanfiction authors the chance to have their work seen by more people, including professional television writers. However, authors quickly discovered that not only did FanLib retain the rights to submissions but also that the site's leadership planned to monetize traffic without compensating authors. Although FanLib received a large number of early submissions, the outcry from a community comfortable with the reciprocity of a gift economy conclusively damaged the company's reputation.[16] Video start-up Crunchyroll faced similar resistance in 2008 from anime fans when its pitch involved profiting off the aggregation of "fansubbed" (fan-translated) content.[17] While, unlike FanLib, Crunchyroll navigated past the community outcry by securing legal distribution agreements for officially subtitled content, the case showed that companies prefer to translate a community's interpersonal connections into monetizable "user-generated content." Recognizing the inevitability of corporate influence, Richard Barbrook claims that the online economy is closer to a "mixed economy," where the commodity and gift relations "are not just in conflict with each other, but also co-exist in symbiosis."[18] Paul Booth refers to this mixed economy as the "Digi-Gratis Economy," wherein mutually productive arrangements between corporations and fans are carved out.[19] Conversely, in her critique of multi-platform content, Suzanne Scott states that companies have perfected a mixed economy that "obscures its commercial imperatives through a calculated adoption of fandom's gift economy, its sense of community, and the promise of participation." This strategy promises to "simply give fans more—more 'free' content, more access to the show's creative team." But it actually constructs a corporate-controlled siloing effect that "equates consumption and canonical mastery with community."[20]

The tension between these viewpoints is central to Social TV. Fans still shared experiences and content in the spirit of a gift-based system, even within the confines of a corporate platform like Twitter or Story Sync. They also eagerly participated in explicitly commodity-driven systems, where they were compensated for and exploited by their "affective labor."[21] Media companies have drastically improved promotional strategies that frame "free" content as gifts, where access to gifts leads to further consumption of cultural products. Yet, unlike FanLib or Crunchyroll, the executives and venture capitalists behind check-ins understood, temporarily, some of the potential pitfalls in appropriating fan activities.

Check-ins did not only disturb gift economies or function like an exchange economy in the guise of a gift economy. Instead, they offered real commodities as gifts, creating what I call a *reward economy*. In these reward economies, the terms of the exchange were stated from the outset. Whereas FanLib and Crunchyroll tried to capitalize on fan labor in exchange for wobbly benefits like "exposure," check-ins promised surer rewards for basic social productivity. And while most of the rewards were centered on ancillary content, check-ins offered up rewards at the point of, and plainly for, consumption. Check-ins' reward economies proved to be a somewhat mutually beneficial relationship for both parties, with users accepting the frictionless sharing and data collection inherent to all platforms to access rewards. In creating collaborative blogs detailing the best maneuvers to exploit the check-in platforms, users also managed to disrupt the stated terms of the agreement to obtain more rewards. Thus, check-ins represented less a type of exploitation and more of a proposition where users could, if they wanted, game the game.

To illustrate these developments, I contrast the two visions of the check-in. Focusing on press reports, executive commentary, publicity materials, and archived activity, I pair GetGlue and Miso's efforts to mirror the aesthetics and functionality of Facebook. In their similar visions, GetGlue and Miso imagined a Social TV defined by conversation, community, and connectivity among fans. The companies later introduced features that encouraged more transient productivity but embedded those features into existing discourses about the value of fan-driven spaces. Yet, these platforms also presumed that fans sought rewards and wanted to display those rewards as a form of subcultural/fan capital. Viggle, meanwhile, envisioned a check-in defined by individualized consumption and consumer goods-as-rewards, without the emphasis on fandom. Viggle explicitly embraced its role as a tool for sponsors, puncturing the façade of Social TV as a fan-centric experience. This comparison shows that GetGlue and Miso's fan-forward operations primarily functioned similarly to Viggle's more expressly capitalistic ad-driven tool. This analysis uncovers familiar attempts to monetize second-screen experiences and the social dimensions of television. The companies aimed to collect user data, to reinforce the value of live viewing, and to partner with an array of corporations—all the while positioning themselves as valuable assets to both the established industry players and the eager Social TV participant.

This approach did not work. The chapter reflects on the failure of check-ins through anecdotes from Andrew Seroff, a former Miso and Viggle employee. Testimony from one professional does not speak for an entire industry, and my intent is not to exaggerate Seroff's observations about the progression of check-ins. However, when combined with the celebratory discourses in the press, this commentary crystallizes how the companies tried to break through the clutter in the Social TV marketplace. The interview also helps clarify that, while a reward-based system presents appeal, it must also offer *enough of a reward* to keep people engaged, or sufficiently interested to leave other social platforms or two-screen experiences.

This chapter continues with a brief survey of digital check-ins and the rise of gamified experiences across industries. Next, I sketch my conception of social productivity and the significance of moments of ephemeral engagement in the Social TV space. The bulk of the chapter is dedicated to my analysis of GetGlue, Miso, and Viggle and their respective evolutions and pivots as the check-in ecosystem faltered. In addressing the struggles of check-ins, this chapter again demonstrates the essential ephemerality of the Social TV era, where content that was once promised to alter the second-screen experience quickly turned fragile or disappeared altogether.

"THE FOURSQUARE FOR [BLANK]": CHECKING IN AND THE GAMIFICATION OF CONSUMPTION

In 2010, digital check-ins were familiar to anyone who had used Foursquare. The app asked people to check in at physical locations to obtain points, badges, and titles (most notably "mayor" of a spot after many check-ins). Foursquare quickly partnered with Zagat, Condé Nast, the *New York Times*, and American Express to deliver special stickers or rewards for checking in at specific locations (e.g., Zagat-approved eateries), or during major events (New York Fashion Week).[22] The company also collaborated with media companies, including promotion for HBO's *How to Make It in America*, which imitated the lifestyle of the program's characters with brand-approved hotspots in New York City. In press for these partnerships, Foursquare was positioned as an ideal platform to market to young, urban consumers under the pretext of food and dining recommendations. As History Channel executive Heather DiRubba told *Ad Age*, though no one knew how to measure

Foursquare activity, media companies saw it as a useful extension of existing social media strategies.[23]

Predictably, Silicon Valley's copycat culture sought to create the next "Foursquare for [blank]." Mo.Pho.to asked users to check in with photos; whrll encouraged them to create "stories" at locations with photos, notes, and tagging of friends; and GoWalla offered them "passport stamps" and trip guides from *National Geographic* and *USA Today*.[24] Even Facebook got into the action with Places, an analogous check-in system integrated into the timeline without the rewards.[25] Others conceived of a check-in beyond physical space. Launched at the South by Southwest festival in spring 2010, Miso positioned itself as a "Foursquare for TV," with digital badge rewards and social networking. GetGlue emerged later in 2010 with a similar round of press coverage and a nearly identical service with one wrinkle: while Foursquare and Miso rewarded users with digital badges, GetGlue mailed stickers to active users.[26] Miso and GetGlue also followed Foursquare by partnering with television networks (HBO, Showtime, PBS) and service providers (DirecTV and AT&T) to guide check-ins during relevant watching periods.

Viggle, meanwhile, borrowed from the features of Shazam, another popular digital media tool. Propelled into prominence in the late 2000s with the advent of the iPhone, Shazam used mobile device microphones to record, analyze, and identify audio clips. Shazam historically has been used to detect songs in public places and on television. But a run of partnerships led to the integration of television series and ads into the "acoustic footprint" recognition software. Combining Shazam-like technology with the check-in methods of Foursquare, Viggle asked users to check in by directing their devices toward the television. Moreover, just as GetGlue had topped Miso's reward scheme by providing physical stickers, Viggle's "loyalty program" took another step toward real compensation with gift cards and consumer goods.[27] Unlike Miso and GetGlue, where users could check in whether they were watching or not, Viggle required that users have content synchronized nearby, and rewarded points based on the duration of check-in.

As this brief history shows, check-ins built their products on the back of established platforms and software. Indeed, rewards, badges, and point systems were also not exclusive to Foursquare, Shazam, or check-ins. Online education programs use digital badges to represent the acquisition of knowledge and skills.[28] Fitness apps likewise reward users with points for meeting their personal health goals. The products exhibit the rise of gamification, where game mechanics (including competition, trophies, and public leaderboards) are appropriated in non-game contexts. In the 2000s, gamification took hold across not just education and healthcare but also the workplace under the guise of improving productivity and efficiency.[29] Health researchers have found that gamification can inspire more effort in participants, but cultural critics have been less impressed.[30] Ian Bogost argues that gamification enabled corporations to "replace real incentives with fictional ones."[31] Gamification fits more sensibly into the entertainment realm, where people expect to be encouraged to consume as much as possible. Still, check-ins framed their incentive programs as decidedly beneficial for participants. In gamifying the viewing experience, check-ins turned the act of watching television into a competitive practice where more viewing—of series, branded content, and ads—led to more points and exclusive benefits.

The focus on rewards also recalls decades of corporate strategy. Networks and brands have collaborated on giveaways and sweepstakes from the outset of broadcasting. Corporate fan clubs encourage diehards to pay for additional access and bonus content. Loyalty programs enable consumers to earn benefits and special pricing when purchasing. Viggle's Chris Stephenson recognized check-ins' place within this history: "Everything that you buy, from coffee to airline tickets, to wherever you're spending money, there's typically a loyalty program in place. It's quite a simple model in that you spend a dollar, the company reserves a penny and packages that up as a loyalty program and delivers it back to you in some cool, brand-relevant way."[32] In framing Viggle as part of this "simple" yet "cool" process, Stephenson promised that his product *gave back* to users. Viggle believed that watching television should not be any different than buying coffee or airline tickets—loyalty deserved to be rewarded.

FROM TEXTUAL TO SOCIAL PRODUCTIVITY

The different strategies used to legitimize checking in as a viable part of Social TV recall scholarly conversations about fan productivity. Responding to the stigmas surrounding fan culture, Fiske argues that fans are "particularly productive" consumers who use their knowledge and taste to counteract a lack of cultural capital.[33] To elucidate this point, Fiske introduces three categories: semiotic, enunciative, and textual productivity. Semiotic productivity is "essentially interior," meaning how all audiences understand and synthesize media.[34] Semiotic productivity transforms into enunciative productivity when fans discuss their interpretations and generate their style (hair, makeup, clothes) to assert membership within a specific community. Finally, textual productivity sees fans create their own "high" value products and circulate them beyond an immediate social group. Naturally, the internet has made fan productivity more visible and more complex. As Cornell Sandvoss and Suzanne Scott respectively assert, modern forms of enunciative productivity extend beyond talk within tight-knit, closed social spheres.[35] Message boards and fan wikis, for instance, are conversational but open to those outside the group. The degree of circulation, then, matters less when most of the activity is public and even searchable. A single tweet can make the enunciative process more public than Fiske ever imagined.

We must recognize that part of conversational, enunciative productivity involves textual production, including the making and sharing of GIFs, videos, and memes. We must also admit that the Social TV experience included activities that fall outside the enunciative or textual categories—in what Booth calls a "liminal" space—like the playing of games or answering trivia questions.[36] It is, therefore, best to work from a new category endemic to social media, one I call social productivity. Between enunciative and textual productivity, users scroll through timelines and feeds, like/react to content, and play games. Social productivity is inspired by Vincent Miller's "phatic" communication, the brief communicative gestures that are "distinctly social but not intended to transmit substantial information."[37] This productivity functions as part of Ethan Tussey's "procrastination economy," as users act within the media industries' "attempts to colonize the daydreaming and multitasking that often takes place as we watch television."[38] But the model accepts that multiple screens and multitasking do not diminish the potential of user productivity. It also assumes that social productivity involves managing—producing, sharing, commenting on—forms of transitory media. This productivity occurs quickly, and often. Finally, the framework acknowledges that the activity often will be public, with the potential wide circulation of material beyond an individual's immediate social circle, but that users do engage in similar activities in private digital settings.

I imagine that this categorization could take evanescent social activity beyond fandom. Conflating the activity on GetGlue with more established instances of fan productivity obfuscates the particulars of social productivity. Put simply, not all users who engage in social productivity self-identify as fans. There are still plenty of debates to have about fan-corporate relationships, but fans are not left outside many media spaces. That means that users engage in social productivity within spaces that assume fandom is the norm and where productivity is *built into* the programming. As Geert Lovink writes, productivity is "a software feature" that "invites us to speak out at—but not so much speak to—others."[39] This new reality does not limit the value or scope of fandom but instead points to how the media and tech industries publicly encourage *everyone* to be and act like fans on structured platforms.

GETGLUE: "YOUR APP FOR TV, MOVIES, AND SPORTS"

Like many burgeoning tech companies, GetGlue used press coverage to craft a familiar "disruptive" brand image. GetGlue intended to be the social platform for viewers self-identifying as fans, even with Twitter (and, to a lesser extent, Facebook) already established in the Social TV space. As CEO Alex Iskold noted soon after GetGlue's introduction:

> Facebook is perceived by people as a great place for friends to connect, see pictures, and send messages. Twitter is admittedly a news network, and Foursquare, GoWalla, and the others are location-based services. It was clear to me that social entertainment was the missing piece, because we're always consuming media.[40]

Two years later, after GetGlue's base had grown to a few million, Iskold remained confident that his company was in "direct competition" with the platforms: "both Twitter and Facebook are very big, but they are completely horizontal. We're focused on content discovery, helping people find what to watch, and connecting them and delighting them."[41] The tenor of Iskold's comments tracked with Silicon Valley norms, where every new product is a solution to or disruption of an antecedent service. Iskold situated GetGlue as a valuable platform that helped people find both like-minded fans and their next favorite series. Facebook and Twitter were conversely ill-equipped to match GetGlue's fan-oriented experience. Iskold leaned on the contrast to Facebook to show that GetGlue provided more fan opportunities: "Checking in is a repetitive behavior that demonstrates continuity. I can like *True Blood* on Facebook, or I can check in to *True Blood* every Sunday night, religiously. It demonstrates I'm a better fan than just someone who Likes."[42] The jab hinted that GetGlue was the space for *committed* fans; people who just "like" television were not real or devoted fans. Iskold evoked the critiques of Facebook's promotion of the like, which, as Carolina Gerlitz and Anne Helmond argue, has distorted affective responses into data.[43] Coverage of GetGlue parroted Iskold's framing. As *Slate*'s June Thomas wrote in her celebration of the product, "there are no barriers to interaction . . . it's much easier to form spontaneous, TV-centric networks than on the big, general-interest social media sites."[44]

GetGlue's interface reflected the competition with the other companies. When users first logged onto GetGlue, the platform greeted them with a sparse white aesthetic and familiar prompt and text box: "Hello, [Name] What are you watching?" Below the "watch box" was another familiar sight:

a real-time feed of friend activity and popular posts from strangers who recently checked into similar series. As with Facebook, profiles displayed photos, friends, and likes, but they also highlighted the stickers amassed by users. Series profiles also provided a Facebook Wall-like space to post media and links. After checking in, users were directed to the live series feed, which highlighted the supposed collective nature of the activity via a tally of the total number of users checked in at a given moment. Perhaps recognizing the limitations of promotional stickers, GetGlue quickly tried to celebrate user chatter already happening on the platform. Introduced in mid-2011, the bluntly named Conversation rewarded engaged users who collaborated with peers and circulated social content. Iskold predictably exaggerated the new addition's impact on fan culture: "The Conversation is smart. It shows all comments from your friends, but only interesting comments from other fans. The Conversation also shows you filtered Tweets, so you can stay on top of what is going on [at] Twitter without leaving GetGlue."[45] While this summary again framed GetGlue as a unique fan experience, it also visibly acknowledged that fans would want to see the broader conversation happening elsewhere.

Users commonly filled Conversation feeds with brief comments that illustrated multitasking real-time viewing: repeating lines, responding to plot points, and delivering jokes. When GetGlue user Tommie checked into a November 2012 episode of *The Walking Dead*, they commented, "Omg. Right after they stab the guy and feed him to the walkers to escape, a KFC commercial comes on talking about 'fresh is better.'"[46] Activity for *The Vampire Diaries* during an October 2013 episode featured a similar number of short comments like "You bet your ass Damon's the FUN brother! XD" and "Why did Jeremy put his shirt back on? :(#Vampnesia."[47]

Still, GetGlue occasionally enabled users to have more detailed talks. User Nunya Business's December 2012 check-in to *Saturday Night Live* included this comment: "Damn ... I knew it. BREAKING NEWS *SNL*: just because you

Doctor Who: Journey to the Centre of the TARDIS

Game of Thrones Season 3 Premiere

Doctor Who: The Bells of Saint John

Doctor Who Season 7B Coming Soon

GetGlue profile circa 2012 displaying number of check-ins, followers, and recent collected stickers.

have a Black host does NOT mean you have to then "ghettofy" the show ... He's an actor ... Not just a BLACK actor." Nunya's comment garnered 24 likes and spurred on a short debate of both *SNL* and host Jamie Foxx.[48] Though the post did not generate complex political discourse, sharing a critique of *SNL* goes beyond Fiske's definition of simple semiotic productivity. Users also posed questions to direct the conversation toward plot points, such as when user candibug76 asked other users posted the question "So is it safe to assume that Ashley doesn't know who gave her all that money?" alongside her check-in to a November 2012 episode of *Revenge*.[49] This query illustrated the social productivity found on check-ins: it received just fifteen responses, a fraction of Twitter or Facebook's sprawling threads, but it did represent a real-time social engagement with content on both screens.

GetGlue imagined a Social TV realm where social productivity brought people together and structured its rewards to inspire that kind of activity. The chief prize was the digital sticker designed with series' logos and promotional photos. GetGlue applied a sticker to a user's profile for each check-in, and, after repeated activity, mailed physical stickers to them. More check-ins and more stickers enabled users to graduate to exclusive titles: Season Fan, Super Fan, Diehard Fan, Elite Fan, and Guru.[50] Users acquired the titles after hitting certain thresholds—five check-ins made one a Season Fan, ten a Super Fan, and so on. But they could only obtain the final prestigious title of Guru by writing "high-quality reviews" that were liked by the community. GetGlue described the Guru as follows:

> [It] is awarded to the user who has the richest level of interaction for the given item based on a point system. Each item can only have one Guru, so you'll need to work for the honor and be vigilant: other eager users can steal the title at any time! The Guru is identified by a point system that rewards high-quality reviews and interaction level. Once you review an item, you are eligible to be the Guru. Points are then awarded when other members of the GetGlue community "Like" your review as well as for your interactions with the item.[51]

Here, the lines between enunciative and textual productivity blurred. Similar to real-time pithy comments, reviews were mostly concise episode summaries. User Marq's review of a *New Girl* episode exemplified this style: "Fun Halloween episode, centering around Schmidt coming to realize his ongoing correspondence with Michael Keaton has all been a lie. Jess as Batman was pretty damn hilarious. Even moreso [*sic*] was Winston as David Letterman (the resemblance was uncanny)."[52] These conversational posts were barely textually productive in the Fiskeian sense. They did not directly address particular users but rather were positioned within an ongoing conversation about *New Girl* visible to other fans. Naturally, GetGlue pitched reviews as *more* than checking in; "especially passionate," "richest level of interaction," "honor," and "vigilant" reinforced the alleged dedication and productivity required to become a Guru. Still, the inflated discourse elided that reviews, like most GetGlue activity, often scored just a few likes or replies.

GetGlue strategized that the visualization of these digital rewards would convince users to be productive. User profiles prominently displayed earned stickers and titles, which appeared next to usernames in a dark color contrasting with the white interface. When users gained a special sticker or upgraded to a new title, their followers were notified of the upgrade. This vision of fandom presumed a desire to discuss as well as an interest in proving one's tastes and expertise. The nebulous appraisal of "high-quality reviews" showed that GetGlue deployed the communal principles of fandom to legitimize the platform. While GetGlue implied that only user consensus could bestow the prized Guru title, it concurrently encouraged the same users to compete with one another for the honor. This contradictory combination of community and competition enabled GetGlue to attract users interested in either component.

GetGlue's corporatized fandom manifested in the controlled distribution of stickers as well. Though the company's main selling proposition centered on rewarding fans with stickers, not *every* check-in generated a sticker. Instead, GetGlue promoted "Limited Stickers" that users could acquire by checking in live and to series produced by network partners. If users checked in to *Scandal* during its primetime airing on a Thursday, they would receive a sticker. Yet, if they missed the live airing and watched on Sunday, the sticker was no longer available. Predictably, then, GetGlue's reward policy did not apply to *every* live episode. When I checked in to a live episode of Fox's low-rated comedy *Ben and Kate* in late 2012, I received no sticker. But *Ben and Kate*'s more popular peer *New Girl* did have a sticker that night, and for every episode that fall.[53]

The temporally restrictive promotion did not stop there. GetGlue offered stickers in the weeks leading up to series or film premieres. Stickers released in summer 2011 promoted the upcoming *Captain America: The First Avenger* with the banner "In Theaters July 22." Users could only obtain the sticker before the film hit theaters. Other Limited Stickers represented the platform's corporate sponsorships. Those for the reality competition series *The Glee Project* were "brought to you by Verizon and our friends at

Oxygen."⁵⁴ To expand the sponsor integration, GetGlue encouraged networks to operate "verified" accounts and sponsored digital badges. It also partnered with Gap and *Entertainment Weekly* to offer a 40 percent off sticker for users that checked in to one of the new fall series the magazine deemed worth watching.⁵⁵ Access to these stickers meant accepting—and displaying—sponcon on the way to Super Fandom.

GetGlue thus tethered its rewards to specific projects and periods, an approach familiar to many Social TV initiatives. First, the approach upheld live television as appointment viewing. Second, it embraced GetGlue's potential for sponsored content integration. Third, it used the popularity of projects to drive attention to the emergent GetGlue platform. These goals were especially relevant given that GetGlue and its partners wanted the stickers to circulate around the web to promote new episodes and products and convince nonusers to join the fun.

In emphasizing visible ranking and reward systems, GetGlue promoted fandom as a fracas over subcultural capital. Borrowing from Pierre Bourdieu's work on cultural capital, Sarah Thornton claims that subcultural capital is a form of currency defined by "hipness." For Thornton, members of subcultures—*ad hoc* taste cultures opposing the "mainstream"—present a sense of "being in the know" through objects or embodiment. Subcultures "put a premium on the 'second nature' of their knowledges," in that no member should be seen trying too hard to be hip.⁵⁶ Thornton's perspective on capital is similar to that of Fiske, who argues that fans produce and negotiate "self-acquired" cultural capital in the realm of popular culture, outside the legitimated arena of "high culture."⁵⁷ While both Thornton and Fiske grant that fan capital is not easily converted into economic capital, the former asserts that hipness serves as a form of distinction. Fans do not procure knowledge solely for enjoyment; they debate, make rules, and jockey for influence like any subculture.

Notwithstanding its discourses about community, GetGlue mutated the principles of fan capital in its user experience. Hipness, obtained through extensive consumption, was the primary currency. Recall Iskold's commentary about how "real" fans should behave. But rather than members defining their capital structures, GetGlue developed the currency with points, titles, and stickers and passed it off as a fan project. This corporate currency was also driven by vague terms to keep users consuming and clicking. Though users could help peers become Gurus by liking their reviews, there was no indication of how many reviews or likes, or what type of analysis, were needed to score the honor. All prospective Gurus knew is that they had to generate "high quality" reviews and have "the most points" related to a particular series. Thus, the most valued rewards in the GetGlue ecosystem were shaped not by a user's commitment or passion but rather by the platform's obscured rules and guided by its corporate partnerships.

GetGlue's manipulation of fan capital gave the company a reason to celebrate. The platform grew from 500,000 users in July 2010 to over one million in July 2011 once it expanded into the United Kingdom.[58] GetGlue also began to release its own weekly and monthly ratings reports, detailing the series with the most check-ins. By late 2011, GetGlue was averaging more than 16 million check-ins per month.[59] About 100,000 check-ins were being shared to Facebook and Twitter every day, inspiring GetGlue to tout that its daily average of "social impressions" (the number of times social content is seen) topped 80 million.[60] Much like Nielsen and Twitter's reporting on the most social series, GetGlue used these charts to validate the company among the public and the media industries. The weekly rankings were reported as news, and studios and networks trumpeted their popularity among fans on GetGlue. For instance, in 2012, Warner Bros. TV celebrated "the real community" of *Big Bang Theory* fans that helped the series score the most check-ins in the history of GetGlue.[61]

By mid-2012, GetGlue's user base had grown to over three million.[62] Iskold gushed about the productivity of GetGlue's users' sharing of real-time chatter and reaction GIFs, as a "sort of the 'next-gen' of fan expression [that the company hoped to place] front and center."[63] Of course, labeling social productivity as both "fan" expression and "next-gen" aimed to set GetGlue as an innovative company. Iskold's reference to "front and center" served as a public promise to current and future users that their productivity would be appreciated and rewarded. Social productivity thus operated as a critical promotional function for the corporations with which GetGlue was partnering. As Iskold said,

> if you're checking in to *Mad Men*, that check-in becomes an ad for *Mad Men*. And when these messages flow down the networks, it's coming from people you trust and follow, not from a banner ad, so you're more likely to click on it. And because of that, the networks and big brands are excited about the way that these targeted messages can travel through the social channels.[64]

Iskold invoked "trust," contrasting it with the banner ad, something most consumers despise. GetGlue proposed to deliver *good ads*, unlike other social platforms or websites. Iskold's excitement about the power of word of mouth signaled that, amid support for social productivity, companies still viewed their platforms as vehicles for promotion. Nevertheless, GetGlue's plan to foster a group of tastemakers did not thrive as much as it hoped. In 2013, the company disabled the review-centric Guru system and pivoted to an even more obtuse points system.[65] This move to a less communal scheme only highlighted that GetGlue, like all social platforms, shielded essential functions from even its most active members.

GetGlue's 2013 update included a new "hyper-personalized" guide featuring live episodes and streaming options and personalized recommendations.[66] Amid the fan-focused hype, the guide was another feature cribbed from more popular companies—and another effort to collect user data to be sold to companies like Gnip, which brokered deals between social platforms and other industries. Gnip CEO Chris Moody called GetGlue a "rich new source of information that will be incredibly valuable to networks, producers, advertisers, and movie studios."[67] The "hyper-personalized" experience, then, stressed GetGlue's aim to assimilate into the media industries. GetGlue made these corporate ties more overt by growing sponsored content and limiting benefits—including ending physical stickers—until it was acquired and later shuttered by i.TV in late 2013.[68] Though GetGlue stressed the conversational facets of social productivity, competitor Miso took a different slant in soliciting more "next-gen" content.

MISO: "SOCIAL TV BEYOND BADGES"

Like GetGlue, Miso launched with venture capital funding and press hype around its Foursquare-like engine. And like GetGlue, Miso promised a fan-oriented experience with check-ins, badge rewards, and basic social productivity.[69] Yet, amid similarities to GetGlue and market saturation, Miso quickly fell behind the competition. In response, Miso introduced new features to take Social TV beyond the check-in, and to capture the attention of more prominent players. Launched in 2010 with the third season of the TNT drama *Leverage*, Miso's Fan Club enabled checked-in users to access bonus content and mini-games related to the series.[70] The name Fan Club was a blatant attempt to construct a community around paratextual content that could be found all over the web. Indeed, Miso curtailed the productive potential of Fan Club by ironically celebrating the skewed, gamified reward system. Ignoring the existing points and badge systems, Miso announced that users would unlock Fan Club bonuses after "somewhere between 1 and 1,000,000" check-ins, and rules would differ by series.[71] Even granting a modicum of humor on Miso's part, the declaration underscored that check-in companies believed that the combination of the fan label and some competitive gamification would inspire deep user engagement.

By mid-2011, Fan Club had floundered, and Miso was losing ground to the better-funded GetGlue. CEO Somrat Niyogi publicly presented his *next* vision for Miso, and the future of Social TV: "Check-ins isn't it.... It's just a starting point.... We don't think badges and stickers are in the long run the reason that people come back." Instead, Niyogi argued, users "want to share specific information about a specific episode."[72] In response, Miso introduced SideShows, a product celebrating more explicit forms of ephemeral social productivity. As the name implies, SideShows asked users to create multimedia slideshows of bonus content that would synchronize to new episodes. Users could fill SideShows "cards" with character histories, trivia, details about diegetic music and costumes, and where to buy them.[73] Similar to AMC's Story Sync, Miso stressed that the second screen should deliver additive information for viewers, not distract with chatter. Miso also promised that users with DirecTV and AT&T U-Verse service would be able to synchronize their SideShows to the live episodes in real time.[74]

Although Miso stated that network representatives would create SideShows content, its pitch centered on the productivity of users, now labeled super-fans and hosts. As Niyogi noted, "TV is an art form, in which every single decision is well thought through. What makes it so great is that people are at the center of its creativity. We believe Social TV should be the same, and we thought these experiences would best be crowdsourced, rather than produced by us, a tech company."[75] Niyogi's remark worked in two ways. First, he linked the "art form" and "creativity" of television to the effort of users, implying that user productivity could be just as crucial to the Miso experience as the televisual content. Second, he tapped into the rhetoric of crowdsourcing by expressing that users, collectively, could make *better* SideShows because of their commitment and knowledge. As José van Dijck and David Nieborg assert, "Mass creativity, peer-production, and co-creation apparently warrant the erasure of the distinction between collective (nonmarket, public) and commercial (market, private) modes of production, as well as between producers and consumers."[76] This erasure is intensified by corporations that need collective labor to generate content and attention for their products. Users become creative producers—as long as they produce within frameworks coordinated by the platform. Predictably, then, in the same announcement, Niyogi offered a fickle vision of Sideshows and user productivity. He noted that SideShows would be a great tool because it could "bring you this information on your phone in a more passive way ... a lean-back way to engage with content."⁷⁷ Will Straw argues that media often produce "particular tensions between stasis or mobility."⁷⁸ Social TV ephemera are especially guilty of this tension, promising at once to inspire more profound, lean-forward productivity and more comfortable, lean-backward consumption.

Still, SideShows functioned how Niyogi promised. User Chad Elkins's SideShow for the *Falling Skies* 2012 season finale began with a poll asking about viewers' reactions heading into the final episode. It later linked to actor Matt Fewer's Internet Movie Database page when the *Max Headroom* star appeared on-screen and offered historical context to the episode's reference to an American Revolution regiment. At the end of the hour, Elkins's SideShow honored a character who died in the episode.⁷⁹ The SideShow epitomized a more robust social productivity where users created content while watching for others to enjoy during multiple airings of the episode. But according to former Miso employee Seroff, the company saw clear distinctions between consuming and creating SideShows: "You were either watching the shows or creating the content."⁸⁰ Seroff said that as part of his contracted position, he was tasked with creating content for SideShows, or "evangelizing" for the product.⁸¹ His account recalls Mirko Schäfer's claim that "explicit productivity" occurs when fans actively develop content, while "implicit productivity" only requires them to interact with preprogrammed interfaces.⁸² Miso sold SideShows as a place for explicit productivity, but, according to Seroff, users operated more implicitly, mostly as observers to the *literal work* of those paid to make the platform seem full of passionate fans.

Despite the framing as a "social entertainment game," for explicitly productive users, Miso struggled to gain traction.⁸³ The user base grew from 250,000 to just 350,000 between 2011 and 2012.⁸⁴ Seroff stated that Miso's attempts to utilize partnerships within the industry eventually limited the potential expansion of SideShows. Once Miso partnered with DirecTV to make it technologically possible to synchronize the SideShows app to a set-top box, competitor DISH Network refused to join the project. Miso also collaborated with Showtime and *Dexter* to produce SideShows content. In doing so, Seroff said that Showtime's rival, HBO, "didn't want to touch us."⁸⁵ As a result, he stressed, Miso created a product that only a "very small group of people" could enjoy as it was intended.⁸⁶ These struggles exhibited the marginalized position start-ups can find themselves in, where partnerships can be as restrictive as they are beneficial.

As a final attempt to improve its standing, Miso unveiled Quips, an app that asked users to share episode screenshots overlaid with commentary, in October 2012. Quips launched with more than 150,000 images for over

500 series, with new photos added each week.[87] In a blog post announcing the product, Miso claimed that Quips

> incorporates the social activity you love with incredibly beautiful images from the very shows you watch. . . . Quips makes you feel like you are sitting right next to your friends on the couch when that hilarious moment on TV comes on [and] you're all pointing to the screen saying, "Can you believe that just happened?!" Maybe you watched the show with them. Maybe you didn't. It doesn't matter because, with Quips, you can start a conversation about almost anything you watch at any time.[88]

Here, Miso curtailed allusions to productivity and returned to the communal aspects of viewing. In a video for Quips, Miso celebrated that the product enabled users to "start a conversation about TV, anytime, anywhere."[89] Productivity and conversation were inseparable; the creation of ephemeral material existed to generate more user chatter. Seroff noted that Quips was supposed to solve the problem of limited productivity seen on the failed SideShows. "We wanted everyone to be authors as well as readers . . . all at the same time."[90] Thus, in responding to its prior exaggerated claims about the scope of user productivity, Miso again invoked familiar tropes of fandom in hopes of inspiring *any* productivity. This last shift recognized that social productivity synthesized conversation, creation, and circulation, but it also signaled that Miso was desperate to find a publicity strategy that effectively sold its products.

The output on Quips exhibited a variety of approaches to social productivity. Some used the platform to discuss the technical aspects of production; as one user's Quip on a *Homeland* image noted, "They [production] are really overplaying the cut to Mike conversing with Chris. Effective though."[91] Some users approached the platform as a space for pithy reproductions of famous lines such as "Say my name" overlaid on a *Breaking Bad* image.[92] Others embodied a snarkier voice; one commented that "I use [sic] to drop acid over there" on a *Fringe* image with a character looking off-screen.[93] Finally, some users made distasteful jokes like "I told you I ran out of fortune cookies" over the image of Glenn, an Asian American character on *The Walking Dead*.[94] Quips content thus served as what Graeme Turner calls productive multimedia "talk" that enabled users to participate in live television's "sense of community or belonging," despite that the platform did not facilitate extensive discussion.[95]

Unfortunately for Miso, Quips met the same fate as Fan Club and SideShows. When Quips launched, the company's executives refused to tell reporters how they generated the live episode images so quickly. While business developer Prakash Venkataraman led *TechCrunch* to believe that Miso had "some patent application stuff" and negotiations with the networks in progress, Seroff told me that was not the case.[96] He explained that once Miso split from DirecTV and AT&T, it could not synchronize its apps to set-top boxes. Instead, it partook in a "highly, highly illegal" practice of "download[ing] every single new episode of live TV at once" and ripping screenshots from every five seconds of those episodes.[97] This tactic enabled Miso to have a night's episodes ripped and ready for Quips during the east coast airing. Miso's "super hacky" strategy helped Quips function, but the delayed rollout stirred too many users to move on from Miso by the time the new product was finally released.[98]

Miso and the SideShows product were acquired by Dijit, another Social TV company, in February 2013.[99] Seroff noted that most of Miso's employees came from a tech background and did not view any of the products as "an opportunity to shape the television industry."[100] Yet, Miso showed the downside of Silicon Valley's eagerness to pivot at the first sign of trouble. Miso correctly assumed that the check-in bubble was soon to pop but also repeatedly shifted strategies without offering users a valid reason to access the platform. Though attempts to build something that appealed to productive users relied on familiar affective rhetoric, Miso did not present users with any tangible rewards in exchange. Users could still gain badges for checking in but only if a check-in was accompanied by new activity on SideShows or Quips. As Seroff put it, smaller companies like Miso "have to bribe people to be active. And you have to have so, so many people be active to be interesting to a big player in the entertainment space. And you have to get that deal [with companies within the media industries] to be interesting enough for people to tell their friends about you."[101] With a lesser user base and fewer rewards to stand out, Miso's fan-centric discourses were even more inflated relative to the industry, but its embrace of social productivity was not successful. Another prominent check-in company, Viggle, took Social TV and productivity in a divergent direction and found eager users awaiting it.

VIGGLE: "YOUR LOYALTY PROGRAM FOR TV"

Launched in January 2012, Viggle embraced its role as a "loyalty program for TV," without the condition that users create or share social content. Executive Chris Stephenson delivered a familiar remediated sales pitch based on Viggle's efficient, intelligent technology:

I really think we're going to change the way people watch TV.... The consumer just taps the app, and in a short period, in five seconds, we take a sample and send it back to the back-end servers.... We can do this if you're watching live or on DVR—and once we know that you're watching a show, we give you points. As you accumulate those points, you can convert them for great rewards.[102]

The Viggle promise was simple: "Watch TV. Get Rewards."[103] *Just tap the app*, and you are on your way to great rewards. Users were not hailed as fans but instead identified as consumers, underlining that Viggle operated more like a loyalty card than Facebook, or another check-in. As Stephenson said, "This is about getting *real* rewards—this isn't about badges and status and all of that stuff. You'll see gift cards for Starbucks, iTunes, Fandango, Amazon, Best Buy, brands that people love."[104] While GetGlue and Miso framed their rewards programs as part of a more profound platform experience, they also celebrated the social productivity that enabled users to obtain those rewards. In contrast, Viggle emphasized "real" rewards from beloved brands and not "all that stuff"—like quickly sharing or liking peers' content.

As part of this exchange economy product, Viggle embraced what van Dijck calls the "triangular relationship" between media companies, advertisers, and consumers."[105] Stephenson gushed about Viggle's ability to bring the three parties together:

> Think of this as like an iAd, that's probably the best way to think about it. We believe that this particular space in the app is really the Holy Grail. We will allow people to check into commercials as well as TV shows. The idea is that I check in and the coupon or the interactive experience is right there on my device at that moment.... When a big brand speaks to the networks about what they're doing with us and says, "We want these ads, and we want you to help us create engagement around the ads that are running on your shows"—we think it's a huge opportunity.[106]

Though Stephenson alluded to an "interactive experience," he referred not to user-to-user activity, but user-to-ad/sponsor activity. Viggle's benefits were exclusively bound to what users could obtain materially in the form of discounts, coupons, and goods. The connection between users and sponsored content made Viggle worthy of the Holy Grail moniker. Viggle president Greg Consiglio likewise noted that "connecting the dots between TV brand advertising and a person's consumer choice" was Viggle's defining feature.[107]

Viggle reward portal circa 2012 showing bonus point opportunities for viewing of a McDonald's ad and featured rewards, including a Kindle Fire HD.

Viggle's exchange-based approach manifested in all parts of the app, beginning with the opening instructions to users: "Tap here to check in. Viggle listens to your TV. You get a point per minute. Redeem points for rewards. Turn on your TV." The specifics of Viggle's point system were relatively clear: users scored one point for each minute of check-in, and even if they switched to different episodes within a given hour, they still received sixty points. But to obtain most rewards, users needed to check in for *many* hours of television. Reward tiers ranged from one thousand points (contest entries to win tickets to events and concerts) to four million points (a Royal Caribbean cruise). A standard reward like ten-dollar gift cards to iTunes or Best Buy required eighteen thousand points. That would require *three hundred hours* of check-in time. Ease of check-in and transparency in rules, then, did not automatically make Viggle a fan-friendly product or make up for relatively small rewards.

To confront this challenge, Viggle encouraged users to score bonus points in even more controlled settings. During my use of Viggle in 2012 and 2013, I scored bonuses for check-ins to popular programs with social media buzz like *The Walking Dead* and *Once Upon a Time*, as well as the first episode of NBC's heavily publicized *Dracula* series. But I received no bonus for my check-in to the post-hype final season of *Gossip Girl*. I also gained point bonuses for merely using the app as a customer of DirecTV, a Viggle partner.[108] And though I could score points for watching via DVR, my point bonuses were given only during live viewership. There was no explicit rationale given for these check-in bonuses. Still, it was unsurprising to see bonuses for more popular series and live activity. Viggle users like me were not greeted with invitations to connect with fellow fans but rather bribed with bonus points for watching pop-up ads from sponsors like Gillette. Stephenson imagined a world where networks and sponsors would participate in a "real-time bidding" to load up their content with bonus points, essentially paying for the privilege to attract users.[109] Furthermore, while program bonuses were restricted to precise periods, ad bonuses were not. Meaning, it was easier to score larger chunks of points by watching ads than it was by watching television. Here again, the path to check-in rewards required users to play by tightly controlled terms and consume additional sponsored messages.

The triangulation of users, content, and ads continued once users accessed individual programs on Viggle's personalized guide. These pages highlighted series through recognizable key art and promotional posters, but directly under those images were bright white banners that visually popped against Viggle's purple aesthetic: "+5 when you watch this"—another ad. Below the banner was another prompt, asking users to visit a program's online store with a Viggle discount code. For example, the portal for *Late Night with Jimmy Fallon* page featured an "NBC Store—15% Off" banner. The focus on online shopping highlighted how easily Viggle integrated into the multiscreen and multitasked viewing environment. Viggle assumed that users could watch on one screen and shop on another, all the while they accumulated check-in points.

Amid the focus on "real" rewards and sponsored bonuses, Viggle occasionally promoted forms of social productivity. The most social product was Viggle Live, a live trivia contest, which launched with the 2012 NCAA men's basketball tournament and later used for the 2014 FIFA World Cup and encouraged users to score bonuses by showing off their sports knowledge.[110] Similar to Story Sync's pseudo-competitive Moral Judgments, Viggle Live required a baseline of attentiveness from users, who had interacted with two screens to answer the rapid-fire questions successfully. But the more relevant parallel to Story Sync is that Viggle's trivia game was also fully individualized, with no visible competition among users. Even in a real-time two-screen realm, Viggle wanted users siloed into a limited experience of watching and collecting points, separate from the wider Social TV ecosystem. Seroff, who worked for Viggle after Miso shuttered, felt disappointed by the former's apathy toward facilitating even the smallest forms of social productivity or community building: "Viggle users didn't seem to care [about conversation]. They didn't care about anything. They just wanted free shit.... Most of the time, you show people an ad, they're pissed, but these people were like 'Hell yeah! The Clorox ad is worth 500 Viggle points!'"[111]

Thinking back to the triangulation, users could benefit from taking any of these actions; watching bonus ads moved them closer to rewards, and discounts on merchandise were a nice perk. Yet, these activities also drove users to watch more commercials, and each click delivered vital information that Viggle and its partners could use to craft more targeted ad experiences. Stephenson affirmed this strategy as many tech leaders do—by pitching it as beneficial for users:

> Once you're checked into *Modern Family*, now we know you're a *Modern Family* viewer—and we have a certain demo[graphic] around *Modern Family* viewers. And our knowledge of Viggle users also includes their previous check-in history, so we can really target the advertising effectively ... the moment you open the app, you're going to see things that we think are most relevant to you tonight.[112]

To engage with a platform promising mass personalization, users must determine if, as Daniel Chamberlin asserts, "the experience at the interface is worth the tradeoff of having [their] actions and behaviors constantly tracked."[113] It is impossible to know precisely how aware users were of this tradeoff, but they embraced its alleged benefits. In 2012, Viggle declared that the episode reminder was its most popular feature among a growing base of nearly one million users.[114]

Naturally, Viggle's Rewards and Deals portals guided users through an interface where commercialization was unmistakable. Exploring the different reward options felt like a digital catalog: glamour shots of products and promotional codes, brief descriptions, and endless corporate iconography. Viggle distilled viewing into a more passive experience, turning users into couch potatoes whose phones promise to bring them gift cards and headphones. A 2013 commercial stressed how easy it was to use, and be rewarded by, Viggle:

> It rewards you just for watching TV. It's simple. Learn more about what's on, play games, and connect with friends, all while earning points. Here's the best part: You can use your Viggle points for real rewards like gift cards, or you can save your points for even bigger

things like a tablet or headphones, or even a cruise vacation. So there you have it. Now TV is more rewarding with Viggle."¹¹⁵

While the spot encouraged users to play games and connect with others, those experiences were secondary to earning points and, by proxy, goods. The ad again noted how easy it was to obtain points ("*just* for watching TV") and what exceptional items users (not fans) could purchase with those points ("best part," "real rewards"). Another 2013 commercial introduced the world to the "first person in history to ever be a full-time TV watcher." In it, a thirty-something man described how easy it is to earn "real-life rewards," to the point where he quit his job to watch television for a living, leaving his disgruntled wife to handle the labor outside the home.¹¹⁶ Around this time, Viggle amassed more coverage framing the product as *paying* users for their viewing. Headlines like "How Viggle Pays People for Watching TV," "Get Paid to Watch TV," and "Stop Working and Watch TV" positioned Viggle activity as an explicit exchange of labor for compensation—not just rewards.¹¹⁷ Of course, though Viggle users were rewarded for their activity in ways that most social platform users mainly are not, company executives and ads avoided references to "work" or "pay." Christian Fuchs claims that platforms rely on unpaid user activity to attract sponsors, forcing users to suffer "infinite levels of exploitation."¹¹⁸ Viggle thus emphasized rewards as a kind of pseudo-payment to position the platform as fairer and more pro-user, which would, hopefully, incite more productive and lucrative user activity.

Compared to GetGlue and Miso, Viggle made it more apparent in consumer-facing operations that it served corporate partners and potential sponsors just as much, if not more than, users. When Viggle tried to buy GetGlue in late 2012, Viggle president and chief operating officer Consiglio declared, "You could imagine that in the near term we'd be looking at ways to expand our advertising platform into that user base."¹¹⁹ The potential merger was not pitched as beneficial for users but instead an avenue to accumulate more ad revenue. In 2013, after the merger failed, Viggle launched its Audience Network, which executive Kevin Arrix called a "centralized way for clients and media buyers to deliver their ads to multiple second-screen apps."¹²⁰ Viggle embraced its role in assisting corporate "clients," with Arrix not referring to the users who constituted the so-called Audience Network. Seroff stated that a critical difference between Miso and Viggle was the size of the latter's ad sales teams. Viggle, he said, "was constantly making deals. It was an app designed to show you ads—and make you like it. The business was ads, and business was good."¹²¹ Seroff said that the partnerships also influenced Viggle content, like when Clorox made a deal with ABC for *The*

Bachelor and then paid Viggle to develop an interactive game that cross-promoted both the series and the sponsor.[122]

Viggle's admission of desire to attract advertisers above all else made it a partial outlier among other check-ins, and among all social platforms that promote themselves as definitively user- or fan-oriented. Viggle and its executives sporadically employed Social TV discourse about collective and connected fandom but more regularly promised a transparent exchange: check in and get rewards. Viggle did not claim to remediate the television experience or try to compete with Facebook and Twitter. This is likely why Viggle remains operational—after its 2015 purchase by another digital rewards company Perk.com—with more than 10 million registered users and its peers GetGlue and Miso have fallen by the wayside.[123]

CONCLUSION: FROM THE GIFT ECONOMY TO THE REWARD ECONOMY

By 2014, networks and studios no longer regarded check-ins as essential to their Social TV portfolios. Miso folded, GetGlue became tvtag and then flamed out, and Viggle diminished in industry reporting despite its endurance. As media critic Simon Dumenco wrote, check-ins were part of a "discrete and linear" logic that assumed the only way to facilitate Social TV success was to prompt an immediate live tune-in.[124] While larger platforms like Twitter could inspire this immediacy on a global scale, check-ins only appealed to a small cadre of users. With networks busy promoting live-tweeting or branded two-screen apps, check-ins were no longer valuable.

Considerations of value were central to the lifespan of check-ins. GetGlue, Miso, and Viggle all tried to sell potential users and partners on the literal value of their visions of fandom and reward programs. Prospective users supposedly found value in the accumulation of points, titles, stickers, or consumer goods. Still, check-ins tried to exploit the increase in multi-screen activity and ephemeral social productivity by suggesting that their respective platforms were *more rewarding* than Facebook and Twitter. Rather than promoting an exclusively affecting and participatory experience, check-ins asserted that users deserved to be compensated for the conventional processes of consumption. Yet, in crafting these reward-driven systems, GetGlue, Miso, and Viggle intended to do what social platforms do: facilitate and collect user activity and generate revenue from that activity. For potential check-in partners, value allegedly came via eyeballs and access to new data streams. Despite the alleged unique promotional discourses and plans here, the fundamental economic models of the internet essentially required that check-ins enact similar profit-seeking strategies.

With their concentration on rewards, check-ins also served as another referendum on the digital gift economy. The reward economy framework grants that corporations set the terms of the exchange but also that users can navigate around these terms as well. For instance, Seroff revealed that the most cunning Viggle users strategically saved up their points until the first of the month, the day when the company would restock its available rewards cache.[125] Similarly, though GetGlue and Viggle created rewards tiers with cryptic objectives, select users worked together on other platforms to create field guides that explained how to hack the byzantine points and rewards systems. A thread on *Sticker FAQ*, a GetGlue-focused site, explained to inquiring users how to space out their check-ins to score the most points. The Vigglers Twitter account, meanwhile, regularly updated users on upcoming limited-time bonuses, like a Dove spot that garnered one thousand additional points.[126] Savvy Viggle users also collected audio samples of episodes and ads and posted them on popular audio hosting website Soundcloud, enabling peers to score bonuses without synchronizing to the television.[127] In this regard, check-in users remixed "collective intelligence," pooling together their resources and knowledge to acquire more reward capital as much as to deepen their platform experience.[128] They managed to circumvent the rules structuring their experience to obtain as many points or rewards as possible.

The failure of check-ins suggests that reward-driven economies may not be the most effective way to inspire long-term productivity among users. Although we can identify some of the false promises of the rhetoric typical to Social TV and broader participatory culture, that rhetoric has proven to be successful in drawing people to many other websites, platforms, and activities. There is an ambivalence toward the algorithmic and data collecting processes of social platforms. Most users have implicitly decided that platform access is worth the exchange of personal information. Indeed, that check-ins could not compete with the enormous user bases of Twitter or Facebook was ultimately the most limiting factor to their reward-driven strategy. Users may have valued the stickers or points they received in exchange for brief moments of productivity but still likely wanted to explore the chatter in more accessible spaces. This is the challenge inherent to a reward-driven system; rewards must be perceived to be valuable enough that users are willing to ignore other platforms to get them. In contrast, Twitter and Facebook do not need to promise rewards. The platforms themselves—where other friends and fans are present and ready to chat or share memes—are the reward.

However, companies believe that modern users see value in the reward economy and check-ins. Companies like Telfie and TV Time have emerged with familiar check-in processes, digital sticker reward offers, and, expectedly, network partnerships. Though less fixated on rewards, Letterboxd has grown into a popular community for people to track and rank their film viewing and connect with fellow cinephiles. The perseverance of this model indicates that the media industries think users can continuously be lured onto new platforms with as little as a digital token. Despite rhetoric about fandom, community, and productivity, corporations still view the user as an easily swayed consumer above all else. The reward economy, then, is just another discursive tool to inspire entry into structured and sponsored digital enclosures.

Chapter 4

"GREAT SHOWS, THANKS TO YOU"

Fansourcing and Legitimation in Amazon's Pilot Season

On February 6, 2014, Amazon Studios, the nascent production arm of digital retail giant Amazon.com, announced the second round of Pilot Season, a showcase for its original television series. Like its competitors in streaming video, Netflix and Hulu, Amazon Studios invested in original productions to supplement its growing library of licensed content as part of the Prime subscription service. But Amazon Studios argued that, unlike the competition online and across Hollywood, it offered a uniquely "transparent" development process driven by "collaboration" between industry professionals, aspirant creatives, and active users of its web community.[1] At its inception in 2010, the studio encouraged prospective filmmakers to submit works-in-progress—from logline pitches and storyboards to scripts and short films—to receive notes from Hollywood executives and regular folks. While the press critiqued this process as an "outsourcing" or "spec labor" gimmick, studio executives consistently praised the "significant" "power of the people," whose feedback acted as a "helpful indicator of what is working and what is not."[2] As a banner on the studio's first website declared, "We invite the audience in early."[3]

Pilot Season emerged as Amazon Studios shifted to television development in 2013. The studio made the first episodes of new projects available for free on Prime Video for roughly a month and encouraged interested parties to watch, review, and spread the word on social media. The evaluation process included brief questionnaires, as well as standard Amazon-style star ratings and space for free-response reviews. Studio chief Roy Price pledged that viewer feedback would play a "very influential" role in determining which pilots graduated to ongoing series.[4] Across the web, the studio celebrated the power of feedback with a provocative slogan: "Watch the Shows. Call the Shots."

However, despite proclaiming to invert the typical Hollywood methods, even Amazon Studios's earliest productions featured recognizable industry professionals in front of and behind the camera. The first Pilot Season in 2013 scored significant attention in the press for this allegedly novel approach, but the crop of pilots included only one project from the studio's participatory amateur pipeline. Once the press responded more positively to projects in the 2014 Pilot Season, the studio shifted its promotional messaging to position itself as a prestigious boutique operation. This shift accentuated the branding markers of "quality TV," chiefly the presence of auteur figures, the artistic freedom granted to those auteurs by the studio, and discursive associations with "better" art forms. More important, this shift minimized the focus on viewer influence. In short order, the studio went from accepting the work of aspiring creators to asking for feedback to not considering consumer response much at all.

Amazon Studios was not the first company to initially call for, and then subsequently diminish, consumer input. The media industries have long solicited feedback, from private test screenings and focus groups to public voting for reality series like *American Idol*. More recently, Hollywood has turned to consumers to advocate for projects on social media or to help fund their production and distribution via crowdfunding platforms such as Kickstarter or Indiegogo.[5] Corporations across sectors ask consumers to rate their products and services after usage. As customers, users, and fans, we are told—now more than ever—that our perspective matters.

Drawing on Amazon Studios's array of websites, press releases, social activity, and feedback forms, as well as the press coverage of Pilot Season, this chapter analyzes how the studio aimed to exploit this *feedback culture* to elevate its industry reputation. The calls for feedback were a corporate manipulation of crowdsourcing that I refer to as *fansourcing*, where an expensive promotional blitz that hails consumers so aggressively to conceal the pivot from a collective, open-source, and transparent experience to a more individualized, fixed, and enigmatic experience.[6] The growth of the internet and affordability of digital production tools has led to the popularization of concepts like crowdsourcing but also "produsage," "produser," and "co-creator," all of which underscore the supposedly increased role ordinary consumers have in shaping professional media.[7] Ideally, crowdsourcing offers a democratizing model for large-scale "wisdom of the crowd" participation that leads to problem-solving and meaningful relationships with brands. But Pilot Season illustrates that corporations are remarkably proficient at integrating these values into their promotional campaigns without actually building or sustaining the infrastructure needed for genuine crowdsourcing.[8] Most important, as with many Social TV campaigns, Pilot Season effaced this lack of infrastructure by situating fansourcing as the best or most significant way to express passion in a multi-screen digital environment.

Fansourcing aims to route the activity of participants toward an array of corporate goals. The approach hopes to instill an ownership or *pre-fandom* among participants, inspiring them to rationalize their consumption of cultural products before they reach the masses by evangelizing for those products on- and offline. For instance, Simone Murray argues that New Line Cinema solicited fan feedback before the release of the *Lord of the Rings* films to minimize unrest among book readers that the adaptation would not meet their expectations.[9] The approach aims to use participant enthusiasm to generate a swell of interest, whether through traditional publicity campaigns or more "spreadable" awareness on social media, among other potential participants and the Hollywood sphere.[10] For example, Will Brooker explores how Sci-Fi and ABC requested fan-produced edits of video footage from *Battlestar Galactica* and *Lost* and then integrated the best products into official promotional materials.[11] In the Pilot Season context, calls for feedback tried to generate attention for the nascent Amazon Studios brand and foster an immediacy in the on-demand streaming setting. Just as conventional networks relied on live-tweeting and multi-screen experiences to produce real-time chatter and social synchronicity, Amazon Studios used fansourcing (and a restrictive distribution window) to manufacture an ephemeral Social TV event and to drive interest for content with no live timeslot.

In framing fansourcing as an empowering way to participate in the circulation of cultural products, corporations secure from consumers quick, free labor as well as relevant data for future usage, without detailing the real impact of participant activity. To better understand how participants would, as Amazon Studios claimed, "call the shots," I provided feedback to three projects during the early 2014 Pilot Season that required little time or room for extensive critique. My feedback represented a form of *ephemeral labor* that has grown considerably online.[12] Brooker and Mark Andrejevic respectively argue that most online labor is "casual, undemanding," requiring "little commitment or effort," and "relatively minor."[13] But beyond the minimal effort, this activity is also regularly 1) restricted within specific temporal windows, 2) immediately inaccessible once submitted to corporate databases, and 3) defined by its ambiguous effect on the processes or content to which it supposedly responds.

While this may recall Maurizio Lazzarato's "immaterial labor," with collaborators working outside of exchange economies, ephemeral labor underlines that this activity is structured, contained, and processed by corporate power.[14] It also stresses the chasm between how companies publicly valorize this labor—again, often as fandom—and what they do with it. For Amazon, the studio remains but a small piece of a global behemoth that dominates online retail, manufactures mobile and home devices, and owns other media companies—all on the back of algorithmic, personalized recommendations. Pilot Season served as just another method to frame data collection and personalization as beneficial for consumers. The company's charge that valuable feedback could be easily distilled into a familiar five-star rating scale and a few Likert-style questions worked to convince participants of Pilot Season's empirical legitimacy. Media corporations have tried to build measurement tools that produce legible depictions of audience consumption since the early twentieth century. From box-office receipts to Nielsen ratings, these "market feedback technologies" promise rationalized measurement but also enable industries to negotiate what constitutes a success or failure.[15] Empirical measures similarly operate as discursive tools of legitimation, whereby the existence of data "prove" consumer tastes. To an extent, then, Pilot Season gave participatory cachet to a process that, essentially, functioned as an algorithmic remix of creating programs to sell products to valued demographics. Indeed, after the Pilot Season project *Transparent* scored industry and critical acclaim, CEO Jeff Bezos identified television content as a new way to get Prime subscribers to shop more: "When we win a Golden Globe, it helps us sell more shoes."[16]

Unsurprisingly, though Amazon Studios claimed that participants would select which pilots would succeed, my participation, and wider discrepancies between feedback and the studio's decisions, proved public opinion to be insignificant. That the early promises about disrupting Hollywood through fansourcing did not reveal the influence of that participation, and that those promises were replaced by markers of quality TV demonstrates that the studio—and much of Hollywood—deemed feedback as a useful stunt in the industry's attention economy. To this end, the chapter reveals that promotional practices driven by quality are more useful for the media industries than those centered on participation because they are effortlessly folded into critical reception and trade press chatter. As previous chapters have examined, the industry press played a central role in creating narratives about tech industry disruptions, uneasiness within the television industry, and collaborations between Hollywood and Silicon Valley. But the studio's pivot to the haven of quality TV exhibits that the press is just as willing to embed new companies into old discourses, especially when those companies legitimize television in the process. This leaves fansourcing participants as potential consumers, fans, or brand advocates but not as significant collaborators.

The following two sections trace the early stages of Amazon Studios, the first Pilot Season experiment, and the participatory hoopla surrounding the supposedly empowering feedback process. From there, I detail my personal experience partaking in the second Pilot Season, which, despite significant hype mostly resulted in brief post-screening surveys and prompts to share on social media. The later sections of the chapter show how Amazon Studios embraced quality TV discourses at the first sign of critical acclaim and industrial legitimation—and thus revealed a disinterest in true or meaningful collaboration with the audience.

"POWER OF THE PEOPLE": THE ORIGINS OF AMAZON STUDIOS

Before it had a $6 billion budget, Amazon's video subsidiary experienced many growing pains. Launched in 2006 to compete with Apple's iTunes store, the then-Amazon Unbox allowed customers to rent or download film and television series on their digital devices.[17] In 2008, the renamed Amazon Video On Demand began offering streaming access alongside rentals and downloads.[18] Amazon changed course yet again in 2011, rebranding the video portal as Amazon Instant Video with a more extensive streaming library for Prime members.[19] The move tried to exploit the vulnerability of Netflix, which faced a backlash for splitting into two companies, one for its streaming library and one for its DVD rental library.[20] To compete with Netflix, Amazon signed deals with CBS and Epix, a licensor of theatrical film releases on premium cable (and a former Netflix partner) bringing the Prime library to more than 25,000 titles.[21] The deals coincided with the release of the Kindle Fire, a competitor to Apple's popular iPad. Together, the moves exhibited the growing power of Amazon corporate integration; Prime customers could watch their favorite films and series on Amazon-branded devices that were shipped in two days.

Amid its rivalries with fellow insurgents Netflix and Apple, Amazon founded a studio to take on Hollywood's entrenched powers. Amazon Studios emerged in November 2010 with an open-sourced film development competition where amateur filmmakers could submit scripts and "test movies" of at least 70 minutes to be assessed, and, potentially, earn a chunk of a preliminary $2.7 million budget. Promotion for the competition underlined the studio's contradictory positioning as a Hollywood outsider that still had access to the right people in Hollywood. An early video identified Amazon as the "movie studio of the future" promising to solve a critical industry problem identified by head executive Roy Price: "Today, the movie business is organized and decisions are made pretty much in one place, Hollywood. At Amazon Studios, we hope to discover voices that might not otherwise

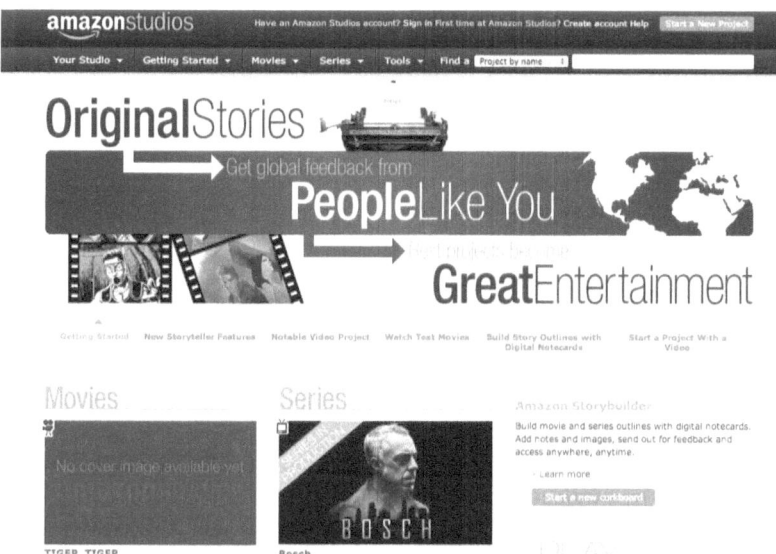

Amazon Studios's website in early 2014 emphasizing the need for "global feedback from people like you" to create "great entertainment."

be heard."[22] A related press release extolled how "film fans" would not only review projects online and through in-person test screenings but also by uploading "alternative, revised versions."

Yet, the press release likewise emphasized the studio's alliance with Warner Bros.—which retained the rights to produce winning scripts or test films—and industry panelists like *Top Gun* screenwriter Jack Epps Jr. The industry connections sold the legitimacy of Amazon's new undertaking to fans, amateur filmmakers, and professionals. While Amazon promised that the competition would "open the doors to Hollywood" for participants, it also aimed to do the same for the new studio.[23] Indeed, the competition exposed immediate inequities in Amazon Studios's priorities. Though fans could provide as much feedback as they wanted, the studio stated that experts like Epps would make the final decisions. From the outset, Amazon Studios promised influence to collective fan action that, in effect, did not exist; authority remained with traditional Hollywood decision-makers.

Unsurprisingly, then, Amazon Studios's lobbying for feedback was less a Hollywood outsider disruption and more of a piece with years of industry procedure. Studios use test screenings to workshop projects based on representative audience response. Test screenings can involve quantitative or qualitative measures, including dials to gauge real-time emotional reactions and more holistic post-screening questionnaires. Test screenings are a natural part of the postproduction process.²⁴ Still, the mythology surrounding them is used discursively within industry circles to prefigure public reception to projects-in-progress. During events like the Television Critics Association (TCA) Press Tour, executives and reporters feature pilots that "test through the roof" and are thus primed for broader success.²⁵ The trades also report on poor test screenings and ensuing revisions, which allegedly signal production trouble that some projects cannot escape upon official release.²⁶ On both ends of the spectrum, collective audience feedback is quantified, abstracted, and used as a publicity tool. Beyond test screenings, the media industries regularly claim to value the potential skill of amateurs. While Hollywood has long operated script contests and fellowship programs to discover new talent, it has also adapted these programs into reality competition series like *Project Greenlight*, *On the Lot*, and *Face Off*. In promising to improve the work of artists through rigorous scrutiny from industry veterans, these series conjure a narrative about the professionalization of amateurs. They guarantee that amateurs offer fresh alternatives to Hollywood's rigid formulas but also that the acumen of industry veterans is needed to evaluate these alternatives. The tension between amateur creativity and professional expertise only confirms Hollywood as a place for wonder and meritocracy, where the hardest-working talents can make the transition from amateur to professional.

Early on, Amazon Studios received sporadic attention for its alleged fan-first practices, including sharp critiques from the industry establishment. Headlines calling the process a "Bastardization of Crowdsourcing" underlined the conflict between amateur and professional voices.²⁷ Infamous Hollywood reporter Nikki Finke questioned feedback's impact on writing credits and compensation. Citing criticism from screenwriters John August and Craig Mazin, Finke called the studio's contractual language "confounding" and warned of exploitation.²⁸ In its first year, the studio evaluated over 6,000 screenplays and 600 test films and distributed more than half of the $2.7 million award funds. But the studio did not commit to financing any project, and it subsequently revamped the scope of feedback due to complaints from entrants. The changes allowed submitters to designate access to their work as "open," "closed," or "revisable by permission."²⁹ Industry panelists expressed relief in this shift, leaning on familiar ideas about the trouble of market testing and the wisdom of crowds. Epps explained that test audiences rarely represent general interest because "nobody knows what they want until it's there." Price admitted that early user feedback was limited because those who enjoy reading and critiquing scripts did not represent the audience that parent company Amazon truly wanted to pursue: "people who like to shop a lot online."³⁰ Yet, a new promotional video saw Price assure participants that "we still believe in collaboration and customer feedback is critical to our development process."³¹ This time, Price tellingly referred to "customers" instead of "film fans," downgrading their efforts from detailed revisions to simple "feedback." This vacillating between specifics and generalities would define the studio's perspective on feedback in subsequent years.

"WHAT GETS MADE IS UP TO YOU": EARLY PILOT SEASON PROMOTIONAL DISCOURSES

When Amazon unveiled the rebranded Prime Instant Video portal in February 2011, the press identified it as the latest rival to Netflix, which had recently surpassed 20 million subscribers and invested $100 million in its first prestige drama series, *House of Cards*.³² Netflix CEO Reed Hastings fanned the competitive flames by declaring that Amazon's approach to streaming and content production was "a confusing mess."³³ Amazon Studios did not push the rivalry with Netflix but did embrace television production to better position the updated video portal. This shift only inspired more comparative reporting revealing stark reputational differences between the companies. *Variety* declared that Netflix had the "creative community swooning" over expensive originals like *House of Cards*.³⁴ In between praising the political drama's star-studded cast and auteur David Fincher behind the camera, the *New York Times* made another industry connection: "It's also the first to be considered . . . as prestigious as the programs on HBO and other top-tier cable channels."³⁵ Conversely, Amazon Studios's declaration about open-sourced television development was situated as an "ambitious experiment" in the face of "some competitive pressure from Netflix."³⁶ Price believed that his studio could move faster through the television development process, given that he hired executives away from Twentieth Century Fox and Sony. "We explicitly decided to include both an open online process of submitting ideas and have a robust traditional development process, reaching out to top talent, top producers," he said.³⁷ Whereas Price once critiqued Hollywood for its monopolistic control over production, he now accepted the need to attract "top" creative talent from the primary industry pipeline.

Despite trying to integrate into Hollywood structures, Amazon Studios publicly embraced its underdog status by wavering back to amateur participation. The studio restated interest in script submissions with the promise of a $10,000 fee to any optioned idea and $55,000 (plus residuals) to any fully produced idea. For Price, the renewed interest in feedback would "defy conventional wisdom" and get "the best answer" about the quality of projects in development.[38] Price evoked one of television's most honored executives—Brandon Tartikoff, who guided NBC through a resurgence in the 1980s—to sell his cooperative process: "We're not so much looking for the next Brandon Tartikoff, but we're all Brandon Tartikoffs, not as individuals but collectively."[39] Price situated feedback as beneficial to the creative life of projects, stressed that the goal of Pilot Season was to satiate consumer tastes, and conjured a utopian vision of collective intelligence: "The connectivity of the web and the reduced costs of producing films create opportunities for people to do so much more in terms of creating entertainment and sharing their ideas and getting feedback. There are millions of people out there who are really interested in movies and TV and who are eager to look at new things and share their opinions."[40] These comments exhibited the studio's central anxieties. On the one hand, Price hoped to sell the prestige of his company's work through a link to television royalty like Tartikoff. On the other hand, he continued to claim that the wisdom of crowds would surpass the expertise of an industry titan and improve professionally produced media.

Producers of early Pilot Season hopefuls, Michael London (*Betas*) and Joey Soloway (*Transparent*), asserted industry support for the participatory process. In a press release noting the inclusion of *Betas* in the first Pilot Season, London said, "Amazon is giving us a chance to work outside the TV bureaucracies and connect directly with audiences hungry for original content."[41] Soloway offered similar sentiments: "In the past when I've made pilots, there's always this phantom testing. This is really a way for people to see it and decide if they like it for themselves."[42] Like Price, London and Soloway strategically pitted typical industry protocol against "the people," exemplifying what Andrejevic calls the "marketing of interactivity."[43] In the producers' minds, Pilot Season acted as a democratic alternative to Hollywood "bureaucracies" and "phantom testing." While London and Soloway's industry experience explained their appreciation of the studio's approach, their comments also highlighted that Pilot Season was mostly creating a new space for existing professionals, not for aspiring amateurs.

To negotiate the ongoing amateur-professional divide, Amazon Studios stressed that those giving feedback were the *right kind of fans* who would have a meaningful role in deciding what pilots went forward to series. "I encourage people to stir up their fans and bring them to the site," Price said during the first Pilot Season. "Those are the people we want sampling the pilots."[44] Price also explained that the studio would "not be as simple as *American Idol*," where the most votes necessitated success. Instead, he affirmed, "You've got to pay attention to the people who have a real passion for a show. In the on-demand environment, that's what matters. People have to reach out for the show."[45] The executive's comments typify a productive view of fandom. Derek Johnson argues that audiences "are not just cultivated as fans, but also invited in, asked to participate in both the world of the television text and the process of its production."[46] Rather than scold fans for their passion, Price put the onus on his team to recognize the passion because it could be productive for the development of Pilot Season projects.

But Amazon Studios remained strategically silent about the precise influence of feedback. Pushed for specifics by the press, Price stuck to a utopian script about making decisions with participants "at our side," conjuring literal closeness between the two groups.[47] Slogans circulated on Pilot Season's website, and social channels likewise declared different levels of influence. The most prominent slogan, "Call the shots," suggested full control, with participants serving as produser-like influencers. A secondary tagline, "You help decide which shows become series," promised a less influential role for participants, who might be able to collaborate with the studio's partners but would not dictate the terms of the process. Finally, "Your opinion matters" assured folks that their feedback would be given rudimentary consideration. The taglines exhibited that, while the studio did not guarantee that participants would sway decisions, it worked hard to hint that they could. Nico Carpentier argues that media companies "work as discursive machines" to promote participation but that they also remain ambiguous to avoid the subject of labor.[48] Companies evade references to labor because it, as John Banks and Sal Humphreys argue, "implies obligation, contracts, and formalized, regulated relationships with producers."[49] Here the studio utilized phrases like "real passion" and "your opinion matters" to pique viewer interest in offering feedback without framing feedback as work.

Beyond the familiar rhetoric, this pitch to participants acknowledged the challenges of divided attention in the on-demand streaming environment. Amazon Studios specifically wanted to attract passionate *fans* with promises of influence in exchange for increased awareness of its nascent brand. While attention from the trade press helped to legitimize the studio within the industry, Price recognized that attention from enthusiastic fans could build buzz among a broader base of potential viewers (and Prime customers). The executive's calls "stir up" and "reach out" about favored projects during a window with a vaguely defined endpoint tried to pressure fans to be as passionate *and as visible* as possible. This phrasing suggested that particularly active fans could get what they wanted as long as they proved their fandom in the appropriate ways and at the appropriate times. Naturally, then, the studio assumed a multi-screen engagement with Pilot Season that could turn the individual experience of watching and evaluating into a more spreadable, collective, and fan-driven social media event. The promises of influence to participants were therefore paired with recommendations for how to circulate that influence further. Across its website and social channels, the studio pushed four-part instructions to participants: "Rate. Review. Share. Tweet. #amazonoriginals." These instructions indicated the correct order of operations for participants, as well as underlined that sharing and tweeting were equally as crucial as providing feedback.

It is also worth noting what was missing from Amazon Studios's discourses at this time. Although the company kept promising to find great new talent outside Hollywood, Pilot Season did not live up to that promise. During the first Pilot Season, fourteen projects (eight "primetime sitcoms" and six children's programs) were available, including:

- *Alpha House*, a political satire from esteemed political cartoonist and Emmy winner Garry Trudeau and starring John Goodman
- *Zombieland*, an adaptation of the popular 2009 film feature film
- *Browsers*, a musical comedy set in an online newsroom from award-winning *Daily Show* writer David Javerbaum and *30 Rock* director Don Scardino
- *Dark Minions*, an animated sci-fi comedy voiced by two *Big Bang Theory* actors, Kevin Sussman and John Ross Bowie
- *Those Who Can't*, a sitcom about delinquent teachers discovered through the studio's open-source development system

A few projects involved new talent but most were developed by creatives or performers with significant industry experience. At this stage, the studio tried to bridge the gap between Price's promises of amateur influence and its commitment to Hollywood pros. The press release for the first Pilot Season promised "a mix of Emmy-winning writers, Academy Award nominees, TV stars, as well as newcomers discovered through Amazon Studios's open-door development process." With one sentence, the studio took another step toward legible industry discourses; awards and stars notably came before the "open-door" practices. The only series described in any depth was *Alpha House*, with Trudeau noting his excitement in working with the "innovative and successful" studio while ignoring the open-door development.[50] Amazon Studios continued to hint at some interest in hearing from viewers. But the studio was already sidelining its previous promise to disrupt Hollywood from the outside.

The results of the first Pilot Season only intensified the disconnect between Amazon Studios's public promises and its actions. Price claimed that within a few days, the pilots received "thousands of reviews," 80 percent of which with four- or five-star ratings.[51] Notwithstanding the alleged passion for the pilots, the studio moved forward with a select few, including *Alpha House*, the most promoted and high-profile effort. On the contrary, *Those Who Can't*, a pilot developed through the open-source system, received positive reviews but was not selected to become a full series.[52] The simultaneous celebration of high star ratings and choice to *not* order many of the projects with those ratings indicated that participant feedback would not, on its own, inspire the studio to invest in future episodes.

"YOUR OPINION MATTERS":
EPHEMERAL FEEDBACK IN PILOT SEASON

To understand the feedback publicized by Amazon Studio, including the effort required by participants, I joined the second Pilot Season test group. During this session, ten pilots (five for adults and five for children) were available for viewing and reviewing from February 6 to March 10, 2014. Promoted as produced by "notable" and "award-winning" creators, the pilots included:

- *The After*, a sci-fi serial from *X-Files* creator Chris Carter
- *Bosch*, a police serial based on Michael Connelly's best-selling novel series, written by Eric Overmyer, former producer of *The Wire*
- *Mozart in the Jungle*, a comedy detailing the elite world of competitive orchestras from Oscar-nominated writer Roman Coppola, *Rushmore* star Jason Schwartzman, and *About a Boy* director Paul Weitz
- *Transparent*, a family dramedy about a patriarch making the male-to-female gender transition from *Six Feet Under* writer Joey Soloway and starring *Arrested Development*'s Jeffrey Tambor

The prompts for participation were highlighted on the right side of the screen, under the "Call the shots" and "Your opinion matters" slogans. The site visually accentuated the importance of feedback by presenting the "Rate. Review. Share. Tweet. #amazonoriginals" directive in a sizeable yellow font set against a clean white background. In an attempt to replicate what would have been a typical Pilot Season experience, I watched the pilots in which

I was most interested. I selected *Bosch*, the detective drama based on the popular Michael Connelly book series. The presentation of *Bosch* looked nearly identical to those for Prime Video's library content: a brief logline about the first episode, a cast list, runtime, and the cumulative viewer star rating. The small differences in the presentation were connected to Pilot Season: "An Amazon Original" banner and smaller versions of the familiar "Watch the show. Call the shots." and "Watch. Rate. Review. Tweet." slogans. The webpage also clarified my role in the process: "Tell us what you thought of this show and help us decide the next Amazon Original Series."

After watching the *Bosch* pilot, I turned to the survey, which reinforced the potential influence of my feedback: "Now's your chance to be heard. Which of these Amazon Original Pilots would you like to see as full series?" But the next line stressed exactly how much feedback was requested: "This short survey should take you no more than 5 minutes per show to complete." Substantive feedback could be given in under five minutes, but the proposed time did not match the public refrains about participation. And the survey was indeed short. The first three questions—the only required questions—presented multiple-choice answers focused on how likely viewers were to watch future episodes and recommend the series to others. Responses were formatted on a Likert scale from "excellent" and "definitely will recommend" to "poor" and "definitely will not recommend." The next batch of optional questions allowed for free responses but also intentionally curtailed potential answers in the framing of the question. For example, one query asked me to list the "two or three things" I liked best/least about *Bosch*. I was then prompted to evaluate a list of "aspects" on the excellent-to-poor scale. Aspects for *Bosch* included "the suspense," "the quality of the acting," "the music," "the beginning," "the ending," and "the setting in Los Angeles." Once completing the *Bosch* survey, the website directed me to continue watching and evaluating. I followed through with *The After* and *Transparent*. These surveys closely matched the first. The three introductory (and required) prompts were identical, while the list of "aspects" to evaluate differed from pilot to pilot. At the end of *The After*, I evaluated "the relationship between the 8 characters," "the look and feel of the show," "the setting during an apocalypse," and "the special effects." As with any Amazon purchase, the website asked me to offer my star ratings and to write reviews. I did not provide this additional feedback to keep my Pilot Season experience anonymous, but thousands of others did contribute in this way. In total, I spent about fifteen minutes evaluating the three pilots.

My participation verified that feedback existed as a type of fleeting labor. My work was easy to complete and relatively generic in scope, with little

Bosch Pilot Season feedback survey asking participants to provide a basic evaluation and detail how likely they are to watch future episodes.

room for specific praise or critique. Even if I wanted to exert more effort in my feedback, the format of the questionnaire dictated that I could not do so. Free-response questions still urged me to ponder simplistic best/worst binaries, as well as asked that I limit commentary to two or three points. The list of aspects encompassed a multitude of disconnected categories. Though one might find it easy to assess music or special effects, other aspects like "the look and feel of the show" were far more ambiguous. While one might liken this process to industry focus group testing, it is worth noting that those events typically involve monetary compensation for participation and are often not promoted with zeal to the general public. Instead, the studio's call for feedback was just another example of the "normative and recruited activity" that Mel Stanfill argues the media industries use to extract value from fans.[53] The feedback was yet another avenue to collect consumer input for *all* future projects, television or otherwise. As Ted Striphas argues, "'crowd wisdom' is largely just a stand-in—a placeholder, an *algorithm*—for algorithmic data processing, which is increasingly becoming a private, exclusive, and indeed profitable affair."[54] Price's comments were revealing in this regard:

> We'll sift through the data.... You have simple metrics like how many people watched it and reviewed it, what their average rating was and

what the reviews said substantively. You also have offline focus groups. We have an online panel recruited from big Amazon movie and TV customers called Amazon Preview, where they'll give more in-depth feedback. There's going to be a lot of data and a fair amount of work.[55]

Price first referred to "data," "metrics," and "averaging rating," and positioned the substance of reviews at the end of the elements under consideration. Despite prior celebrations of passionate fandom, here, Price affirmed that participants' role as "data provider[s]" superseded the specifics of their feedback.[56] His description of Amazon Preview and the collective of "recruited" viewers offering "more in-depth feedback" also accepted that Pilot Season feedback was less detailed and less valuable. Price thus continued to speak of the import of feedback to position the studio as a unique operator, even as he admitted to relying on traditional test screening feedback as well.

After three weeks of deliberation, Amazon Studios announced series orders for six pilots from the second Pilot Season. Price praised the breadth and depth of viewer feedback: "We had a tremendous response to Amazon Studios's latest pilots—in fact, double the number of customers watched these pilots compared to our first season, and they posted thousands of heartfelt reviews with pleas for us to continue these shows."[57] Price did not reveal specific details supporting his claims, nor did he clarify the role of the "heartfelt" reviews. The mix of data, qualitative commentary, in-person market testing, and executive observations functioned as a black box, where vapid signifiers about the passionate response elided an internal decision process.

The available data hinted that feedback did not radically impact which pilots were ordered to series. When Amazon Studios began to review the data, *The After* and *Bosch* received the most reviews by far (over 10,000 each). Because Price's comments emphasized passion, it made sense that the raw number of responses helped those projects. However, two other pilots ordered, *Mozart in the Jungle* and *Transparent*, only received around 3,300 and 2,770 reviews, respectively, far fewer than *The Rebels*, the one project *not* ordered to series. This discrepancy suggests that the *quantity* of public response was not the only determining factor. Similar conflicting data points emerged from the user ratings. The order, from highest-rated to lowest-rated, was *Bosch, Mozart in the Jungle, The Rebels, The After,* and *Transparent*. The one pilot not picked to move forward (*The Rebels*) received better reviews than two surviving projects (*The After* and *Transparent*).[58] The result hints that participant evaluation of the *quality* of projects also did not govern studio decisions. Notably, the press release announcing the Pilot Season spotlighted known talent like Chris Carter, Michael Connelly, and Joey Soloway, while the writers/producers of *The Rebels* were not at all mentioned.[59] It appears the studio always had less interest in *The Rebels*, or, more likely, had more interest in working with famous figures. No participant response could shift that interest.

THE ECHO CHAMBER OF QUALITY TV DISCOURSE

While Amazon Studios's promotional discourses inflated the power of feedback, the circulation of those discourses by the press played just as important a role in affirming the studio's claims and positioning it as an industry upstart. Headlines such as "Amazon: Vote for Shows You Want Us to Make" and "Amazon Shows Off Its First Shows, and Wants to Know What You Think" condensed the studio's brand to a utopian collectivist. They also curtailed discussion about the quality of the Pilot Season projects or the waffling between amateur and professional voices. This framing was equally beneficial for the press and the studio. The press situated Pilot Season this way to build its metanarrative about tech companies and streaming portals upending entrenched Hollywood powers. For instance, *Slate* suggestively asked, "can Amazon transform TV?" while *Variety* labeled Price a "disrupter" who was "Rocking the Business of Content Production, Distribution, and Consumption."[60] Rather than underlining the instability of the industry, the press used developments like Pilot Season to establish an agenda about particular parties thriving under uncertain industry conditions. The framing as an old school-new school rivalry also functioned to promote the Hollywood apparatus as innovative and forward-thinking; embracing Amazon Studios as part of that apparatus made Hollywood look better.

Meanwhile, the lack of detailed critique of the first projects enabled Amazon Studios to focus on Pilot Season as a participatory operation, even though the studio had failed with a similar approach to film development a few years prior. The framing of Amazon Studios as a futuristic disruptor helped launder its reputation within the industry, assuring prospective collaborators that the company had found innovative ways to make television. In the first Pilot Season, the studio only ordered five of fourteen projects to series, three of which were about children and thus completely ignored by the press and critics. This outcome did not matter as much as establishing that Amazon Studios was open for business. The aided buzz helped attract higher-profile talent for the second Pilot Season. Unlike the first Pilot Season's affordable live-action and animated comedies, the second edition delivered a mix of comedies and dramas from more recognizable stars—and zero projects submitted through the open-sourced development system. This swing indicated the studio's march toward another, more popular industry discourse framework: quality TV.

Defining quality TV has long been of debate in television studies. Robert J. Thompson labels the "quality TV aesthetic" as comprised of textual conventions, including large ensemble casts, interwoven plotlines, sociocultural awareness, and artistry historically associated with cinema.[61] For Thompson, this combination of conventions positions quality TV as more like a unique television genre. Jane Feuer argues that quality refers to "delivering whatever demographic advertisers seek, or ... attracting an audience with enough disposable income to pay extra for TV."[62] But quality TV is perhaps best understood as a corporate branding strategy. Branding provides the discursive scaffolding that normalizes ideas about particular textual characteristics and specific demographics with both public and industry audiences.[63] For instance, the "It's Not TV, It's HBO" slogan situates HBO against the rest of television's vast wasteland. But HBO's other promotional discourses accentuate formal conventions—narrative complexity, big casts, genius writer-producers, expensive production design—that also seek to appeal to wealthy, educated, and, increasingly, male audiences.[64] As such, HBO's quality-oriented discourses validate individual programs just as much as they legitimize the pay cable model.[65] After an expansion of original series helped solidify HBO in the 1990s, competitors such as Showtime, FX, and AMC have used similar male-fronted series and quality discourses to target valued demographics with allegedly "better" programming.[66] Netflix followed in the streaming realm by outbidding HBO with its $100-million investment in *House of Cards*, which catalyzed a rivalry between the two companies. This blend of quality TV *as a type of programming* and *as a type of discourse* has been historically successful in launching pay cable, basic cable, and streaming video as destinations for "prestigious" or "high end" television.[67]

Again, however, it must be recognized that the press plays a critical role in fortifying quality discourses. Michael Z. Newman and Elana Levine argue that the coverage of industry innovation aims to elevate television to a higher cultural category alongside cinema or literature.[68] Networks and the press work in concert to circulate quality discourses to legitimate one another, and television as an art form. Similarly, Christopher Anderson argues that HBO's emphasis on auteurs and intricate character dramas helped expand interest in television criticism, which, in turn, further solidified an "echo chamber" around HBO as a pillar of prestige.[69] This symbiotic relationship between networks and television critics did not begin with HBO. It did, however, deepen in recent years as the increase in scripted television production on cable (and then streaming portals) coincided with a surge of online writing about television, subsequently paving the way for the continued legitimation of masculinized quality TV discourses.

It is no surprise, then, that the media echo chamber that promoted Amazon Studios's participatory processes would be receptive to the studio's turn toward quality TV. The second Pilot Season's increased commitment to dramatic programming and high-profile talent in front of and behind the camera inspired a more positive response from the press and critics. Instead of headlines highlighting the crowdsourced feedback, the press celebrated the studio's "big step forward," and how the "latest pilots are a cut above its last batch—and most other pilots."[70] The response to *Transparent* was particularly transformative. Critics praised the pilot for its nuanced, uncommon representation of gender identity and transitioning. Citing its "authenticity and specificity," *Slate* critic Willa Paskin called *Transparent* "an honest to goodness great pilot that feels—and I mean this as a compliment—exactly like one of those HBO shows with a 1-to-1 ratio of viewers to think pieces."[71] Recall that among viewers who gave Pilot Season feedback, *Transparent* scored both the fewest public reviews and the worst overall star rating of the five primetime projects. Yet, despite a level of viewer disinterest, *Transparent* was ordered to series. The decision showed that the collective viewer opinion would not invalidate positive press attention. Indeed, the glowing critical response to *Transparent* spurred Amazon Studios to move its branding further away from participatory feedback and closer to quality TV. This shift included an amplified focus on visionary auteurs, the creative freedom granted to productions, and associations with other forms of "high" art. The change in strategy did not entirely erase the influence of viewers on the studio's brand image. Still, it marginalized them to exploit discourses that the press was already beginning to circulate.

"MORE THAN A TELEVISION SHOW": AMAZON STUDIOS'S SHIFTING STRATEGY

Amazon Studios's quality TV discourses began in earnest in March 2014 as it announced the results of the second Pilot Season. The press release included critical praise for *Transparent*, *Bosch*, and other surviving projects, formally embracing the company's newfound industry cachet.[72] Price reiterated critics' admiration for the "very distinctive" *Transparent* and gushed about the "real vision" of creator Soloway. He also said that it "was very important to have [*Bosch*'s] Michael Connelly involved every day in the production and planning and postproduction" and that working with Chris Carter "brings a lot of expectations" from *The X-Files* fan community.[73] This maneuver stressed the prized presence of the "showrunner-auteur," a figure that Newman and Levine argue underlines the artistic status of television as an authored text.[74] Soloway, Connelly, and Carter were positioned as the "distinctive" and "important" voices pulling the studio toward legitimacy.

But in quality TV discourses it is not enough for the showrunner-auteur to be identified as a singular author of a prestige series; they must also be given creative autonomy to craft that vision by respectful and open partners. Michele Hilmes argues that Steven Bochco brought a creative resurgence to network television in the 1980s due to "a greater degree of creative control over the programs than they [writer/producers] had before."[75] Despite Hilmes's analysis of Bochco's work on broadcast, discourses about creative freedom are more common in cable, and with men.[76] Showrunner-auteurs like David Chase (*The Sopranos*) and Kurt Sutter (*Sons of Anarchy*) are encouraged to tell stories not permitted on broadcast airwaves and given extended episode times on the less rigid basic cable schedule. Discourses about autonomy function to establish the showrunner-auteur but also strive to legitimate non-broadcast modes of production, where quality TV contrasts with the mass-produced nature of network television. Naturally, discourses regarding the autonomy granted by Amazon Studios emerged alongside those celebrating showrunner-auteurs like Soloway. At the July 2014 TCA Press Tour, *Transparent* star Jeffrey Tambor praised the superiority of Soloway's work, calling the series "the most transformative experience" of his career, and "all I ever wanted to do as an actor. It reminds me of Broadway." Star Gaby Hoffmann shot down the claim that the cast and crew were exchanging lower pay for artistic freedom: "This notion that we're making some sort of sacrifice for Amazon, it's completely false. I would have done this show for no pay ... It is 100 percent a privilege."[77] Soloway fortified their cast's assertions about autonomy: "This has been absolutely the least amount of interference ... than anything I've ever done."[78] Here, both cast and showrunner-auteur confronted potential criticisms of working with Amazon by stressing the creativity to be unlocked with a respectful partner. In doing so, Tambor, Hoffmann, and Soloway marked *Transparent* as quality TV, solidified the studio as a production destination, and assured other professionals that they too could have a "transformative" experience with no interference.

The studio's reformed discourses also sought to elevate certain pilots closer to more venerated artforms. Soloway praised the decision to release all episodes of *Transparent* at once because they viewed it "like a five-hour movie more than ten episodes." Executive Joe Lewis echoed Soloway's description and verbalized the studio's hope to redefine the television form. "It's novelistic; it's not episodic. We're actually getting to make up this new form of storytelling as we do it," Lewis said. "We need to figure out a new word for it—it's not film, and it's not TV."[79] *Transparent* was distributed through a streaming portal, but it certainly mimicked the format and style of television. Despite the digressive plot, nonlinear storytelling, and directorial flourishes of the series, the narrative still unraveled episodically with familiar mini-arcs along the way. Jay Chandrasekhar, star of the third Pilot Season project *Really*, also evoked this talking point by claiming that the studio had developed an experimental atmosphere akin to "a new independent film scene."[80] The efforts to extricate the studio from "normal" television mirrored HBO's exclusionary "It's Not TV" slogan, which, according to different analyses, conjured art cinema, modernist theatre, or "high culture."[81] Thompson argues that oppositional branding approaches are common to quality TV discourse and "pick up on the old idea of defining quality by what it isn't."[82] In the comments, Amazon Studios projects were novelistic, like independent film and also not at all like film, or entirely new—but not television.

After the 2014 TCA panel, Amazon Studios hustled to shift focus from crowdsourced feedback to discussions of the projects themselves. In an interview about the studio's progress, Price openly identified his competition for the first time: "in terms of high-end half-hours and hours, there are a few places that are on the list. There is a type of show that probably can go to AMC, FX, HBO, Showtime, Netflix or Amazon.... there's a group of people who get a look at a certain show, and we're in that group."[83] Price cleverly linked his studio to the "few places" that are good enough to attract "high-end" programming, and that receive the most acclaim among critics and within the industry; no additional context was needed. Similarly, near *Transparent*'s season one debut in September 2014, Soloway referred to their series as "more than a television show" and, indeed, more akin to "a movement." *Transparent* actor Amy Landecker added that she was attracted to the potential sweeping impact of the project. "It could—and I'm not being hyperbolic—save people's lives," Landecker said.[84] These comments advanced the notion that the studio was dedicated to making art about relevant social issues and presented challenging ideas, characteristics often evoked in quality TV discourses.[85] Notably, no part of the studio's newfound mission attended to the value of viewer response to these relevant and challenging ideas. Once *Transparent* survived Pilot Season, Soloway and their cast stopped talking about how the feedback improved their supposedly important work.

As executives and talent tried to change the conversation about Amazon Studios in the middle of 2014, the trade press willingly allowed those people to speak in quality TV clichés and crafted stories about the new momentum catalyzed by the second Pilot Season. In August, global business outlet *Quartz* penned a story about Amazon Studios "finally challenging Netflix."[86] Similarly, a September *Variety* report identified the studio's new "core principle": "pick the right projects and get out of the way." This story referred to audience feedback as a "pilot-bakeoff" with no mention of the viewers who were previously so important.[87]

By fall 2014, Amazon Studios faced a crossroads as it introduced both the third Pilot Season and the first season of *Transparent*. The new pilots included the comedies *Really* and the Steven Soderbergh-produced *Red Oaks*, along with *The Cosmopolitans*, a dramedy written and directed by Oscar nominee Whit Stillman, and *Hand of God*, a drama about a rogue judge played by Golden Globe winner Ron Perlman. The press release for the third Pilot Season underscored the "passionate and talented" big names in front of and behind the camera before it referenced viewer feedback.[88] On Twitter, the studio highlighted the new prospects and again invited viewers to offer their input, retweeting regular viewers recounting their participation. Yet, once *Transparent*'s first season debuted, the studio's social channels turned full attention to publicizing the wave of positive reception to the series from both critics and Hollywood stars. In choosing to occasionally retweet viewers' affection for giving feedback while also selecting LGBTQ+ stars like Elliot Page and Jane Lynch to vouch for *Transparent*, the studio privileged one type of discourse—and one type of viewer—over another. It also established a sense of quality through celebrities who could speak to authentic and positive minority group representations. At this stage, the Pilot Season website underwent a makeover as well. Though still claiming that participants could "call the shots," the slogan had been visibly minimized on the site. More detailed descriptions of the new projects and their associated auteurs, as well as larger photos of the various stars, took on a new visual prominence. These maneuvers are subtle but telling. They reflected an Amazon Studios that was less concerned with its status within the industry, one that is confident in its ability to select projects that speak to a particular audience, and one that is no longer in need of feedback-driven discourses.

FROM AWARD WINNER TO "GONG SHOW": THE DOWNSIDES OF QUALITY TV

Despite the increased presence of recognizable names, critics did not celebrate the third Pilot Season projects like they had *Transparent*, and the data of public viewer feedback showed similar signs of disengagement. The percentage of four- and five-star reviews compared relatively equally across the February and August 2014 Pilot Season runs, but a smaller number of total reviews hinted at a lack of interest. Within the first month of the February 2014 Pilot Season, participants delivered 30,432 total star rating reviews to the five pilots.[89] In contrast, the five projects in August 2014's Pilot Season only drew 9,613 total star rating reviews across a similar period.[90] To an extent, this disparity arose from the built-in fanbases for February pilots *The After* and *Bosch*, both of which attracted more than 10,000 reviews. But the lack of interest in the projects in the next Pilot Season suggests that the studio's enlarged industry profile did not necessarily generate more viewer feedback. Yet, the two Pilot Seasons were most similar in their final results. Like in prior Pilot Seasons, a project with relatively high interest and reviews from viewers (*Really*) did not survive the process, while a lesser reviewed project with a high-profile name attached (Soderbergh's *Red Oaks*) did.[91] Then, in January 2015, the studio scrapped *The After*, the project with the most enthusiastic viewer response across the first three Pilot Seasons.[92] These decisions proved that passionate fan interest increasingly did not hold much weight.

While Amazon Studios continued to contradict its perspective on feedback, its parent company sought cachet in other ways. In mid-2014, Amazon announced a $300 million agreement to make HBO's library available to stream on Prime Video and available to non-HBO subscribers for the first time.[93] Rather than just borrowing from HBO's playbook, Amazon tried to associate with HBO and its pristine reputation. The deals with other producers of prestige television inspired a new corporate collaboration known as the Streaming Partners Program (later Amazon Channels), which allowed Prime customers to add a la carte subscriptions to HBO, Showtime, Starz, PBS, and 100 other partners. As Michael Paul, VP of digital video noted, this program hoped to integrate the brand value of outside channels into Prime Video. "We make it very clear and lean into the brands of our partners," Paul said. "We think it's really important for users to have the relationship with those brands and those shows."[94] Amazon assured that it would not disrupt the relationship its customers have with HBO or Starz or PBS. It would, on the contrary, facilitate that relationship by enabling new and customer-friendly access points.

Amazon Studios also diversified its offerings by returning to film production and distribution. Though the company began as a disruption of Hollywood cinema, it had not officially produced a film until Spike Lee's 2015 project *Chi-Raq*. As with television, the studio's reinvigorated film work suppressed any reference to new amateur creators and instead chased the prestige of the festival circuit. The majority of the thirty films it produced between 2015 and 2017 were co-distributed with recognizable "indie" companies like Roadside Attractions, Magnolia Pictures, IFC Films, and Annapurna Pictures, and aimed at the Cannes, Sundance, and New York film festivals. In courting indie distributors and filmmakers like Lee, Woody Allen (*Café Society* and *Wonder Wheel*) Kenneth Lonergan (*Manchester by the Sea*), and James Gray (*The Lost City of Z*), the studio situated itself as an auteur-friendly partner uninterested in the modern blockbuster economy. Independent studios and production houses have utilized this tactic before. Alisa Perren traces how Harvey Weinstein's Miramax inspired an independent cinema wave in the 1990s by promoting filmmakers like Steven Soderbergh, Kevin Smith, and Quentin Tarantino as fresh voices who produced "sophisticated material geared toward a more educated and discriminating audience."[95] As Perren argues, Miramax's success was as much predicated on its marketing campaigns as the content or style of the acclaimed projects.[96] Amazon Studios's aping of Weinstein's practices extended to genuine production agreements with his newer production house, the Weinstein Company.[97] Once more, the press positioned the studio as an insurgent figure in Hollywood. Popular movie site *Collider* referred to its practices as "Operating Like a 1970s Studio."[98] A *Variety* profile detailed Price's last-minute attempts to woo Lonergan at Sundance and secure the rights to *Manchester by the Sea*, which would later win two Academy Awards.[99] Amazon Studios did not pitch its film expansion as a critique of its television productions or Pilot Season. Instead, the move was another step toward legitimacy in both realms, accentuating the studio's overall influence.

Amid Amazon Studios's efforts to associate with noted Hollywood emissaries of quality, Pilot Season continued with increasingly vague pitches for feedback. The studio operated four Pilot Seasons in 2015 and two in 2016. Public data is sparse, but results from the January and November 2015 Pilot Seasons showed more discord between viewer interest and final decisions. The seven pilots of the January 2015 group attracted a total of 18,733 star ratings, while the six in the November 2015 pod scored 18,198 star ratings over similar month periods.[100] This was more than the August 2014 edition but still far below the peak in February 2014. When averaged to consider the variance in the number of pilots, interest in Pilot Season had stabilized:

- February 2014: five pilots, an average of 6,086 votes per series
- August 2014: six pilots, an average of 1,922 votes per series
- January 2015: seven pilots, an average of 2,676 votes per series
- November 2015: six pilots, an average of 3,033 votes per series

In both 2015 Pilot Seasons, the studio moved ahead with the top-rated series: *Man in the High Castle*, an adaptation of a famous Philip K. Dick short story, and *Z: The Beginning of Everything*, a fictionalized account of Zelda Fitzgerald's relationship with F. Scott Fitzgerald, also based on a popular book by Therese Anne Fowler. But both Pilot Seasons saw the studio make choices that conflicted with some of the data. In the January group, the studio ordered *The New Yorker Presents*, a documentary series from the venerable magazine, that saw the lowest viewer interest. Meanwhile, in the November group, the studio chose not to move forward with *Edge*, a project from *Lethal Weapon* and *Iron Man 3* writer Shane Black, even though it had nearly double the number of votes than the next closest project. In these cases, connection to some source material—particularly books, which represent a key piece of Amazon's larger enterprise—helped a project succeed within the Pilot Season realm. At the same time, the company did not always acknowledge general fan passion.

At this point, Amazon Studios began to overhaul its development pipeline, allowing high-profile writers, directors, and projects to bypass the Pilot Season process. This shift occurred first with Woody Allen, who the studio paid over $80 million to skip the pilot process to make six episodes of *Crisis in Six Scenes*.[101] To explain this new strategy—a common way to attract A-list talent that does not want to participate in a protracted piloting process—executive Joe Lewis said that "with some people, it's worth it to roll the dice and let the person surprise you."[102] The studio then signed direct-to-series deals with television auteurs, including David E. Kelley, former *Lost* showrunner Carlton Cuse, and *Mad Men* creator Matthew Weiner.[103] Far from its original pitch to disrupt Hollywood practices with undiscovered talent, the studio signaled that the industry's established producers could jump to the front of the line.

Meanwhile, a trio of Pilot Season successes—*Z: The Beginning of Everything*, *Good Girls Revolt*, and *The Last Tycoon*—were quickly canceled after their first seasons arrived on Prime Instant Video.[104] The studio then paid $50 million for the live rights to the NFL's *Thursday Night Football* and $250 million to adapt *The Lord of the Rings* novels.[105] Amazon Studios previously declared that it was "not in the business of being 10 million people's third favorite show, we're in the business of making someone's favorite show."[106] But Price noted that the new goal was to find "big shows that can make a difference around the world."¹⁰⁷ Reporters claimed that the studio was changing course due to Bezos's disappointment with Price's work, but the executive told *Variety* that consumer feedback led him to these changes. "We've been looking at the [viewer] data for some time, and as a team, we're increasingly focused on the impact of the biggest shows. It's pretty evident that it takes big shows to move the needle."¹⁰⁸

But the tide had turned against Price and Amazon Studios among the industry power brokers and journalists it once courted. The trades reported on the emerging perception of a "difficult working environment" at the studio.¹⁰⁹ David E. Kelley, who was given a direct-to-series order for his project *Goliath*, called the studio "a bit of a gong show" with leaders who were "in way over their heads." Though previously praised for its lack of interference, producers who had worked within the Pilot Season structure complained of confusing or delayed notes from decision-makers.¹¹⁰ The press also critiqued the studio for doling out its direct-to-series orders to men but forcing women like Amy Sherman-Palladino, creator of *Gilmore Girls* and the Amazon Studios project *The Marvelous Mrs. Maisel*, to participate in Pilot Season.¹¹¹ In October 2017, Roy Price was fired after facing sexual harassment allegations from Isa Hackett, a *Man in the High Castle* producer, throwing the studio into further turmoil and drawing more attention for its dysfunction.¹¹²

With a burgeoning shift in strategy and Price removed from his post, Amazon Studios unveiled another Pilot Season on November 10, 2017, featuring the work of noted artists like author George Saunders and *Sex and the City* producer Michael Patrick King.¹¹³ Despite the normal nods to feedback, the studio quickly moved on from all three pilots by December, bringing the end of Pilot Season and the calls for participation.¹¹⁴ The studio also canceled prior Pilot Season successes in *One Mississippi*, *Jean-Claude Van Johnson*, and Soloway's second project, *I Love Dick*.¹¹⁵ Outside of the children's programs, most of the projects developed through the Pilot Season process were canceled by 2018. The company also severed ties with Weinstein and Allen after they, too, were accused of—and charged, in the case of Weinstein—sexual harassment and assault.¹¹⁶ As such, Amazon Studios was undone by the desire to achieve industry legitimacy. The studio scored attention by making bold proclamations in the press and agreeing to work with controversial men like Weinstein and Allen in hopes of becoming a Hollywood dominant player. However, these choices moved the studio away from its original vision as a corrective to the traditional media empire. Amazon Studios altogether jettisoned the early emphases of its brand: discovering amateur talent and facilitating meaningful viewer feedback. In making such claims or deals to angle for acclaim, the studio tried to game media industries discourses without following through on those discourses. It quickly shifted from participatory discourses to quality discourses and had no place to turn when its output did not meet industry standards for prestige.

"CALL THE SHOTS?": CONCLUSIONS ON PILOT SEASON

It is difficult to know precisely how Amazon Studios initially considered viewer feedback, or to what degree those considerations changed over time. But it is clear that the studio's requests were vague and that the studio shifted tactics at the first indication of critical acclaim. The shift proposes that Amazon Studios used participatory culture and fansourcing to make a splash within the industry, a strategic move the studio then discarded when more esteemed options emerged. The transition to promotional discourses that were more likely to be circulated by the press raised the studio's profile in a way that viewers "calling the shots" never could. Fansourcing and quality TV are not opposing discursive strategies, but the progression of the studio's promotions shows that these discourses serve distinctive functions and values to Hollywood. The masculinized discourse of quality TV is simply too appealing to the press, as it works to legitimate television as an industry and an art form. Conversely, the industry—from studios to the press to talent—sees fan participation as little more than a gimmick.

This scenario creates a bind for interested consumers. On the one hand, Pilot Season was a small opportunity to see behind the curtain and bypass the existing structures that dictate how media content is developed, evaluated, and distributed. Projects like *Bosch* and *The After* showed that passionate viewer response could, potentially, affect the decisions of a large production house. On the other hand, the studio convinced thousands to watch content and provide ephemeral—but not unmeaningful—labor and then failed to reveal the exact effect of this labor. Participants chose to provide feedback and potentially enjoyed doing so, but even still, it is difficult to see this activity as a genuine form of participation. No one was given particularly notable access to any part of the production process. This ephemeral labor was dictated by the studio, which hoped that freely giving content away would turn viewers into fans and eventually convince them to pay for subsequent episodes via Prime subscriptions. In this regard, Pilot Season was more fraudulent than a usual Hollywood testing screening or a crowdfunding campaign because of the studio's constant insistence that it represented something more disruptive and more meaningful for participants.

Yet, to the credit of viewers everywhere, it is worth underlining that Pilot Season was not a rousing or lasting success. The failure suggests that people were sharp enough to see through the heightened promises about calling the shots. Situated in context with other failed Social TV initiatives, we see a broader picture of consumer skepticism or disinterest in corporate campaigns that emerge with no connection to existing content and sweeping claims about a participatory revolution. Modern fans are very willing to help crowdsource, crowdfund, or simply create a crowd in the name of their favored popular culture artifacts. But fans are also increasingly cognizant of the role that companies want them to play—even if that means just giving a few minutes of their time to fill out a post-viewing survey. Thinking more broadly about feedback culture and the endless array of surveys that offer product discounts in exchange for telling a corporation "how we're doing," Amazon Studios did not provide enough incentive to potential participants to keep Pilot Season afloat. Beyond the lack of transparency in the process, it appears that viewers did not overwhelmingly love most Pilot Season projects, an issue punctuated by the studio's meandering strategies and Price's eventual exit. The studio's struggles show that many Social TV initiatives emphasized the *social* too much and the *TV* too little.

Notwithstanding a self-celebratory discourse of disruption, we should not ignore the recurrence of these strategies. Modern consumers may have more options and access than ever, but corporations piggyback on those evolutions to imply new relationships or opportunities for participation. Social platforms and campaigns like Pilot Season are purportedly situated to rectify the challenges of concentrated media ownership, content clutter, and enigmatic ratings systems. Instead of offering a viewer-focused revolution, Pilot Season and parallel calls for feedback merely shift the mysteries of market testing to a new realm and inventing equally byzantine frameworks that keep most everyone in the dark. The media industries will only continue to employ these kinds of outreach tactics because they are useful to attract the attention of fragmented audiences and the press. Company after company in the Social TV era pitched themselves as more collaborative or fan-friendly simply to inspire more consumer engagement, whether publicly on social media or privately. The former helps turn fansourcing into corporate spreadability, while the latter enables companies to collect data en masse without the hassle of traditional representative sampling. Fully integrated mega-corporations like Amazon can fold research and development costs into other businesses, integrating Pilot Season data into the algorithmic infrastructure that churns out personalized recommendations and inspires a more profound commitment to the Prime subscription ecosystem. Thus, while Pilot Season was an industry failure that also did not fully take off with consumers, the campaign certainly generated intelligence that Amazon will be using for years to come.

Chapter 5

"IT'S WHAT CONNECTS US"

HBO and Platform Authenticity on Twitter

On Sunday, April 20, 2014, at 9:00 p.m. EST, HBO retweeted a reminder about a new episode of *Game of Thrones* from the program's account:

> @GameofThrones: QUIET IN THE REALM. #BreakerofChains starts now on @HBO. Silence your ravens and spread the word. #gameofthrones

HBO's retweet directed followers what to do (pay attention to their screens and remind fellow fans) and when to do it (at that moment). The retweet also shared two hashtags and a photo of the "Breaker of Chains," Daenerys Targaryen (Emilia Clarke), to help the series and related hashtags trend on Twitter. We might expect reminders about upcoming episodes that attempt to convert social media users into television viewers from television networks on Twitter. Other tweets from spring 2014 promised access to live Q&As with the stars of *Silicon Valley* and *Veep* and shared praise for *Last Week Tonight with John Oliver* from celebrities. But they all served as promotion for HBO's existing brand image defined by prestige "It's Not TV" programming.

Three years later, HBO took a different method for building its brand on Twitter. On June 6, 2017, HBO posted a tweet with a GIF from a newer series *Divorce* featuring star Sarah Jessica Parker's Frances exasperatingly stating, "Oh come on." The accompanying text read: "@HBO: When you realize you're not even halfway through the week . . . #HBO #DivorceOnHBO #FYC" With the use of a GIF and the familiar "When you realize" tweet structure, HBO tried to generate a branded meme for social spreadability. Posted at 2:12 p.m. EST on a Tuesday, the tweet's declaration of "not even halfway through the week" and Parker's irritated "Oh come on" referenced the banality of the typical US workweek. The tweet courted bored followers who might be scanning their social feeds for fleeting moments of distraction or entertainment and would be willing to share the post with others facing similar workplace doldrums. Rather than purely promoting impending episodes or a prestige aura, HBO situated itself as a more relatable brand embedded deeper into the lives of followers and fans. Yet, at the same time, HBO's strategic use of hashtag slipped in some publicity for the network (#HBO), the series (#DivorceOnHBO), and a Primetime Emmy Awards campaign (#FYC, which refers to "For Your Consideration" in industry parlance) right before awards ballots were due. Accordingly, HBO's tweet characterized a complex promotional process where social or viral-baiting content appeared beside—and part of—familiar branding strategies.

As previous chapters have demonstrated, Twitter remained a central site of interest for the television industry and Silicon Valley throughout the Social TV era. While live-tweeting and hashtags quickly became standard ways to generate bursts of social attention, synchronized multi-screen apps and reward-based platforms tried to capture viewer enthusiasm and direct it past the Twitter timeline. But by the second half of the 2010s, two developments altered how the dominant players in television approached Twitter. First, the industry's attempts to sustain live and ad-supported primetime programming were further marginalized by an enormous increase in spending on original content by streaming portals. The expansion of streaming options made traditional live broadcasting more obsolete to a growing number of consumers and limited the potential potency of a weekly real-time collective viewing experience. Second, this period also saw a notable evolution for Twitter into a complicated space of sociopolitical tension, where subcultures of wannabe influencers, comedians, reporters, media figures, activists, and hate groups clashed regularly.[1] Corporations, which historically used Twitter for publicity, became willing to exhibit new sides to their "personality" by engaging with trending topics or other users in more pointedly comedic, activist, or hostile ways.[2] As participants in this shifting landscape, television networks experimented with new ways of sociality that surpassed reminders about soon-to-occur live airings or interactive promotional events. Along with more direct engagement with users or the creation of attempted viral content, this newfound sociality divorced brand building from the primetime schedule and integrated it into the rhythms of everyday life.

The decision to show more personality on Twitter or other digital platforms embodies what Sarah Banet-Weiser calls the "authenticity and sincerity" deployed to improve a company's "affective connection" with its potential consumers.[3] While corporations now spread their brand iconography across the web, the amount of content clutter has motivated them to produce and perform what I call *platform authenticity* to overcome the attention deficit. The concept of platform authenticity acknowledges the inherent performativity of the branded corporate voice online. It is a voice devised in creative meetings, executed by multiple people working to appear spontaneous, ironic, or sympathetic, and met with passive acceptance or ambivalence as a reality of the modern media ecosystem. Platform authenticity also recognizes that this corporate brand voice aims to adopt the vernacular, style, and conventions of specific social platforms. While major brands strive to construct a consistent voice across platforms, they likewise modify that voice to best integrate into the conversations and content circulating among regular users. Platform authenticity looks different on Twitter than it does on Instagram, and, as this chapter exhibits, looks different on the same platform as time goes on. This form of social media-oriented authenticity underlines that modern corporations try hard to situate themselves directly alongside target audiences in online spaces, experiencing, and responding to, the world in the same way as those audiences.

For television networks, the performance of platform authenticity works to circumvent the ephemerality of social media and establish not just permanence but *omnipresence*. Instead of pushing viewers to participate in coordinated and controlled live experiences built around television, platform authenticity conditions them to develop an ongoing relationship with networks via a constant barrage of content, including both traditional promotional material and modern memes and political activism. In his exploration of the "procrastination economy," Ethan Tussey identifies the rise of short-form "media snacks" that try to "ensure people can find the flavor, texture, and indulgence appropriate to their circumstances," including multitasked leisure and work boredom.[4] Not every network embraced this turn equally, but the growth of platform authenticity and branded snack content speaks to an evolution of Social TV strategy to account for more screens, personalized timelines, and on-demand schedules and, ultimately, take multi-billion-dollar companies beyond television.

This chapter charts the evolution of platform authenticity. I analyze HBO's Twitter activity across two periods, first in spring 2014 and then in spring 2017. A close reading of the activity shows that, in 2014, HBO was satisfied to celebrate the prestige of its original programming and its ability to impress critics and attract big-name stars. In focusing on original projects, HBO predictably ignored the majority of its licensed film library, reruns, and late-night programs and limited the public visibility of its engagement with non-celebrity users. In this regard, HBO operated on Twitter as it did in most promotional realms, situating itself as an authoritative voice for Hollywood excellence with little need for a hard sell. My survey of the 2017 activity, however, reveals that HBO performed platform authenticity to advance its brand to a new generation of potential subscribers. While HBO did not entirely abandon quality-oriented branding, the network did develop a more spirited tone, replied to followers more often, participated in more non-media trending topics, and pressed to generate easily shareable, viral-baiting content. This strategic performance of online culture and vernacular tried to firmly place HBO more acutely within the screens, timelines, and lives of followers and viewers.

Beyond the broader changes that impact television and social platforms, HBO's pivot to platform authenticity reflected notable changes to its premium subscription business model. HBO long resisted providing à la carte access to non-cable or satellite customers—known as "cord-cutters" or "cord-nevers"—due to lucrative partnerships with cable and satellite providers. But after years of industry speculation about an "unbundled" HBO, and after Netflix prompted a dramatic uptick in spending while promising to "become HBO faster than HBO can become us," HBO executives relented.[5] In late 2014, CEO Richard Plepler declared that it was "time to remove all barriers to those who want HBO" and that the streaming arena "should no longer be left untapped."[6] I argue that the creation of the direct-to-consumer streaming portal, HBO Now, led the network to cultivate a new approach to online branding to capture those "untapped" opportunities.[7] HBO had to sell savvy social media users and cord-cutters/nevers on the value of an HBO Now subscription in a market increasingly dominated by Netflix. To do this, HBO acted as Netflix CEO Reed Hastings predicted it would: by expanding its identity from a prestigious television network to an omnipresent multi-platform content machine.

A few additional points are worth making regarding the method in this chapter. The two periods of analysis featured in hundreds of HBO tweets and retweets, only some of which are addressed here. While the close examination of tweets (or any activity on any platform) poses a challenge regarding the number of texts required to get the most precise picture of tweets and retweets, Twitter's advanced search functions allow for meaningful filtering of accounts, hashtags, and periods.[8] This approach is not comprehensive but does generate a representative view of HBO's evolving social persona across two particular eras. The two periods are also notable because they fell during critical promotional cycles for HBO, featuring new episodes of its highest-profile drama (*Thrones*), comedy (*Veep*), and news/variety show (*Last Week Tonight*). HBO had a significant incentive to tweet and promote at these times.

It is also worth noting that the author of HBO's tweets is unknown. It could be a group of interns, a mid-level member of the marketing team, or an outside boutique agency that specializes in social content. But this question of authorship is valid for most publicity materials and only further muddled by the performed intimacy of social platforms. Yet, I presume a coherence of strategy on HBO's part because the tweets were published with the explicit purpose of articulating an image and voice to the public. All brand material, while the product of many, effaces that collaborative work to present a unified corporate image. As Catherine Johnson argues, "branding becomes, then, a frame through which industry discourse about its own working practices and values is articulated."[9] Thus, despite their ephemeral nature, this chapter treats tweets like any other finished publicity paratext.

Below, I contextualize HBO's platform authenticity within a history of branding by the pay cable behemoth. I then present a cache of HBO tweets from 2014 that illustrate the network's efforts to retain its historically prestigious brand identity via digital and discursive associations with top-tier Hollywood talent, critical acclaim, and minimal conversation with fans. Next, the chapter situates HBO's showdown with Netflix and the announcement of HBO Now as a major fulcrum in the streaming sphere that also required a reformulated approach to social branding. From there, I survey 2017 tweets that demonstrate this shift to platform authenticity, including the celebration of binge-viewing, holidays big and small, and the banality of everyday life—all realms to be occupied by an HBO series or meme.

BRAND BUILDING AT HBO

Though the use of mascots, catchphrases, and colorful labels dates back to the nineteenth century, the idea of what Naomi Klein calls a "corporate consciousness" emerged in America during the post–World War II economic boom.[10] With the expansion of corporate globalization, exploited labor, and ad-supported mass media, companies across sectors decentered products as their intangible "brand essence" and instead stressed what brands "mean to the culture and to people's lives."[11] Branding has also grown exponentially in the media industries. The push into international and ancillary markets has, as Charles Acland notes, made Hollywood content a "mutating global product" scaffolded by "cross-promotional webs."[12] Powered by comprehensive research and vertical and horizontal integration, media conglomerates craft malleable, location-specific campaigns in foreign markets, target niche audience segments within domestic markets, and direct press coverage into future branding opportunities.[13] The pervasive nature of branding—Acland's "permanent marketing campaign"—structures how consumers, critics, and industry professionals consume and discuss media.[14]

In television, branding's role expanded in the 1980s amid mounting media consolidation, the explosion of cable networks, and far-reaching audience segmentation. During this period, cable networks branded themselves by types or genres of programming (the Weather Channel, Food Network) and audience constituencies (Lifetime, Nickelodeon).[15] As programming guides grew and technology provided consumers more control, branding strategy advanced to consider how each series fit an internal "brand filter."[16] Sam Ward argues that network brands have morphed into transmedia-esque entities that no longer exist "solely on television" and instead create meaningful moments across different platforms.[17] This migration to digital, social, and streaming platforms builds a thoroughly branded ecosystem where episodes and promotional clips flow into social updates, games, publicity interviews, discussion forums, and approved fan-generated content, all attempting to translate a unified image to viewers.

Johnson argues that the perceived interactivity of this digital ecosystem is "ideally situated to 'relationship branding,'" with media companies more apt to offer viewers the chance to participate in two-way communication.[18] This vision aligns with Celia Lury's claim that a brand is best understood as an "interface," framing a two-way exchange of information between producers and consumers.[19] Adam Arvidsson likewise argues that a brand serves as a "frame of action," guiding people to develop an "affectively significant relationship" with products and services.[20] For Banet-Weiser, this relationship is similar to one between two people: an "accumulation of memories, emotions, personal narratives, and expectations."[21] In these theorizations, consumers are thought to play an active role in building the relationship—what Arvidsson calls "co-creation" of the experience.[22] But this is not a genuinely collaborative process. Viewers may express loyalty to a network that influences which programs remain on the air, but the agency promised to consumers only generates value for the brand owners. In television, networks leverage any suggestion of a two-way relationship to sustain viewership.

For HBO, a premium cable company, the promotion of an aura of prestige works to sustain this relationship and, significantly, convince subscribers to continue paying the monthly fee on top of the usual programming bundle. Christopher Anderson refers to this process as HBO establishing "a unique *cultural* value" to protect its economic value.[23] Part of this cultural value manifests in HBO's original programming. With fewer series to produce and no advertisers to which to answer, HBO has strategically pushed boundaries in narrative complexity and subject matter while still operating within the conventions of television storytelling. HBO is credited with catalyzing the latest "golden age" of television thanks to series like *The Sopranos*, *The Wire*, and *Sex and the City*.[24] Yet, as described in chapter 4, HBO has also courted press attention to bolster its reputation as a superior network operating above conventional television. The praise heaped upon HBO by the press during the early 2000s played a central role in confirming the prestige constructed by the "It's Not TV. It's HBO" slogan. Thus, HBO also sustains its relationship with consumers by, as Avi Santo argues, promising cultural capital to the subscribers watching the quality programming.[25] Dana Polan calls this the "performance of distinction," where viewers who pick up on the intertextual references or artistic inspirations in HBO series produce a "cultural citizenship" that helps them feel smarter than the average watchers.[26] In this regard, the HBO brand symbolizes Lury's interface concept. "HBO" is a multivalent entity whose meaning is negotiated by the company, critics, and audiences.

However, television network branding is more complicated when imported to social platforms, where, as Banet-Weiser argues, "authenticity itself is the brand."[27] Legacy brands are reconstituted on Twitter or Instagram by invisible employees who construct and maintain the "authentic voice" of a multi-national corporation. Crucially, this authentic voice is a performed and manufactured sensation where people *feel* like a corporation's content speaks to them, perhaps enough to want to, as Nicholas Carah argues, "work together [with the corporation] to create and act out shared values."[28] But for HBO, which has long promoted prestige and cultural capital to subscribers, authenticity meant initially mirroring its approach to branding elsewhere. Rather than embrace the conversational potential of Twitter, HBO first used the platform as another space to circulate publicity images or videos as well as praise from critics and industry professionals. In this way, HBO remained wedded to the existing idea of the "Not TV" brand without much consideration of the social aspects of a social platform.

@AUTEURS AND DISCURSIVE LINKAGES ON TWITTER

HBO's spring 2014 tweets reinforced its aura of quality by focusing on its original programming, the distinguished stars linked to the brand, and laudatory comments from critics and industry professionals. However, HBO spent most of this promotional energy trying to convince people to watch the programming live. These temporal reminders tried to inform and persuade Twitter users to take a simple action: watch the program about to air live in their time zone. Temporal reminders often directly addressed the user and tried to construct a sense of urgency connected to the new episode. For example, precisely as the first episode of *Silicon Valley* debuted on HBO's schedule (April 6 at 10:02 p.m.), HBO tweeted: "@HBO: Big dreams. Corrected vision. Complicated facial hair. @MikeJudge's #SiliconValleyHBO starts now." Santo argues that HBO's branding "no longer strictly conveys a sense of aesthetic criteria ... nor does it identify a particular demographic."[29] Here, HBO eschewed an explanatory plot synopsis for a new series; instead, the abstract promo-speak presumed that followers already *knew* what kind of excellence to expect. The most informative piece of the tweet came through the "mentioning" of *Valley*'s co-creator Mike Judge, the respected voice behind *Beavis and Butt-Head* and *King of the Hill*. Mentioning Judge efficiently told users that *Valley* would present a specific worldview, voice, and style and was worth watching simply for Judge's involvement, as well as accentuating the performed immediacy of the tweet's "starts now" pitch. To further perpetuate the auteur imagery, HBO ignored the two other writers, John Altschuler and Dave Krinsky, who co-developed the series with Judge. The network's existing commitment to what Anderson calls a "more widespread discourse of authorship" extended to the Twittersphere where the digital links between an auteur's project and the network are made literal.[30]

Though HBO sometimes retweeted fan comments in this period, it was more common for the network to share praise from others within Hollywood. On April 13, Mindy Kaling, formerly a writer/star of *The Office* and star of her own series *The Mindy Project*, tweeted: "@MindyKaling: Damn @MikeJudge, *Silicon Valley* is so fucking good. Everyone watch right now." HBO retweeted Kaling's adulation for Judge and *Valley*, creating an additional affirmation loop for the series, Judge, and HBO. That the praise came from Kaling, a budding Hollywood auteur with a passionate online fanbase, underlined the quality of HBO's new comedy. Rather than directly push people to watch the second episode of *Valley*, HBO used Kaling as a tastemaker to increase word-of-mouth interest in the series. The network also retweeted congratulatory praise on April 27, the premiere night of *Last Week Tonight*, which was HBO's answer to *The Daily Show* with former cast member John Oliver. Among those retweets were laudatory and anticipatory updates from actors Gillian Jacobs and Colin Hanks:

@GillianJacobs: The hilarious, smart, nice and bespectacled @iamjohnoliver has a new show-@lastweektonight. It begins tonight. Congratulations John!

@ColinHanks: Very much looking forward to @lastweektonight with the one and only @iamjohnoliver. Ok probably not the only john Oliver, but a GREAT ONE

Judging by the tone of these tweets, Jacobs and Hanks were merely happy for their friend Oliver. But HBO retweeting the actors served as another bit of promotion for a new series, just a few hours before its first episode. Despite HBO's reputation for quality, not all of its series are successful. HBO's sharing of these tweets intended to boost the profile of *Valley* and *Last Week* early in their runs to ensure a sustained level of social media buzz. The tactic signaled to users that HBO programming was *so good* that smart people working within the industry, those making and starring in beloved series elsewhere, turned to HBO for personal entertainment.

Unsurprisingly, given that HBO has routinely routed industry media coverage its publicity materials, HBO used critical discourse to establish its social brand. The functionality of Twitter—links, retweets, hashtags—makes it easier to exploit the "echo chamber" of coverage and publicity and share those materials directly with potentially interested users.[31] For example:

@sepinwall: My interview w/ "True Detective" creator @nicpizzolatto about the end of season 1 and hints of season 2 http://tinyurl.com /kz2udkr (March 9)

@HBO: "Dante, Redemption, and the Last #TrueDetective Essay You Need to Read" by @ComplexMag: http://itsh.bo/1gi3ToJ @McConaughey #WoodyHarrelson (March 31)

@Vulture (*New York Magazine*'s blog): Watch John Oliver stop by The Daily Show to be British and rub in how much better HBO is: http:// vult.re/1jL3HQL (April 25)

These three updates—two retweets by HBO and one HBO tweet—exhibited how the network drew from a broader conversation to underscore its quality. Retweeting prominent US critic Alan Sepinwall and his interview with *True Detective* creator Nic Pizzolatto identified HBO as *the* place for important auteurs. The tweet about the *Complex* essay on *True Detective* was loaded with all sorts of signifiers of quality, referencing the famous poet Dante and linking great literary figures to the HBO drama. These tweets exemplify how, as Polan's insists, HBO "trains" its audience to "take cultural works to be enigmas or puzzles in which one goes beyond the text at hand to something else."[32] Reference to "the Last Essay You Need to Read" implied that people had already read other essays about *Detective* as if that is simply something HBO subscribers do. While the tweet did not mention Pizzolatto, the inclusion of stars Matthew McConaughey and Woody Harrelson ensured that HBO would still be associated with big-name stars.

Meanwhile, the *Vulture* tweet about Oliver's appearance on *The Daily Show* reinforced the perceived "Not TV"-ness of HBO, even if it was done facetiously. Both Oliver's comments in the clip and the tweet jokingly alluded to HBO's greatness despite the apparent similarities between *Last Week Tonight* and *The Daily Show*. Though the former delivers a similar format and tone to the latter, it airs on HBO and is, thus, "better." Polan argues that HBO has consistently mocked the concept of "watercooler" television despite producing programming that fits that designation. He contends that, in HBO promotional material, "High seriousness is overlaid smugly with a knowing wink, a putting of deep purpose into quotations."[33] Even when headlines or tweets comically accept the cultural and industrial hierarchies between HBO and the rest of television, they still preserve those divisions within the discursive sphere. The retweet function is particularly useful here, as HBO slyly spread the reputational contrast in the guise of lighthearted needling from Oliver. HBO was not stating its supremacy outright but instead conveniently sharing the perspective from an employee and a respected news organization.

Retweets enabled more than discursive linkages from outside the brand orbit; they also facilitate corporate synergy that fortified the broader brand image. Paul Grainge defines synergy as "a principle of cross-promotion whereby companies seek to integrate and disseminate their products through a variety of media and consumer channels, enabling 'brands' to travel through an integrated corporate structure."[34] In this case, HBO promoted material from specific series accounts or related programming events. This type of tweeting and retweeting involved more conventional promotional tactics in the guise of behind-the-scenes access to, and tidbits about, the respective series, occurrences, or performers:

@VeepHBO: #Veeple, ask @mrmattwalsh anything during his live @reddit_AMA, going on now: http://itsh.bo/Qeo2q5 (April 11)

@GameofThrones: TODAY AT 12PM ET: Ask questions for a live Q&A with @Maisie_Williams at @HBO Connect. ASK MAISIE: http://itsh .bo/1jv3LW2 #gameofthrones (April 23)

HBO could have tweeted this information from @HBO, but the strategic retweeting of *Thrones* and *Veep* provided a brief bit of synergistic promotion for the series and their accounts. The tactic also allowed the main HBO account to mimic stereotypical affective fan-like vernacular and user engagement without undercutting its prestige brand. The *Thrones* tweet used an all-caps reminder of the Q&A's start time, while "ASK MAISIE" connoted an excitement in chatting with a *Thrones* star. The *Veep* tweet offered a pun-worthy nickname and shareable hashtag with #Veeple and read as if the account was talking directly to users. These were minor but notable moments where the HBO audience was framed as a fan community. Naturally, maneuvers like capitalization and nicknames seek to normalize specific responses and behaviors in ways that Hollywood can exploit—even if that exploitation is just a quick promotion for web Q&As.

Will Brooker's concept of "overflow" is again instructive here. Brooker contends that television programs offer so much textual content that a singular entity cannot contain it, and, thus, the extra content subsequently must "flow" into secondary paratexts.[35] Social platforms magnify the potential for textual overflows but also illustrate how paratextual material flows through multiple platforms, creating cascading overflows. HBO's synergistic tweets introduced new nodes of contact for those interested in particular parts of the brand. Beyond those for individual series, HBO operated accounts for its documentaries (@HBODocs), boxing coverage (@HBOBoxing), online streaming platform (@HBOGO), and press team (@HBOPR). HBO cunningly used the sub-accounts to distribute overflow content for particular audience groups but would then open the flow of content back to the main HBO account when most beneficial—e.g., before the live broadcast of a new documentary. This approach permitted HBO to diversify, synergize, and segment its audience, while also saving the main account from the cluttered push marketing strategies that the audience might not associate with the prestige of HBO.

FACILITATING ENGAGEMENT

HBO's focus on original programming in 2014 also colored how the Twitter account interacted with non-celebrities. Though most theorists agree that brand management requires, as Johnson argues, "enabling and utilizing the input" of consumers, HBO generally avoided publicly partaking in this type of engagement.[36] When HBO did seek minimal forms of public user input, the account ensured that the engagement did not distract from the curated messaging pushing people to watch its Sunday night programming. Indeed, when HBO called for more direct communication with the audience, it directed folks to other spaces like HBO Connect, a website designed for chatter between fans and stars. During this period, the majority of HBO's Sunday night tweets referencing viewers followed the format seen in an April 13 update:

@HBO: Have questions about tonight's #SiliconValleyHBO? Ask them for @Amandaccrew's #HBOConnect Q&A & you may get answered http://itsh.bo/1jz3owC

Here, HBO's account performed a more explicit form of user engagement with its conversational tone and direct mode of address. Still, the tweet functioned more as a one-way broadcast than a legitimate conversation starter. More important, like the two-screen apps detailed in chapter 2, HBO wanted to turn viewer conversation and activity away from Twitter and toward a platform that the network could control. At the time, HBO promoted Connect as the approved "pulse" of social conversation. Connect brought together live Q&As with actors and showrunners, real-time streams of chatter from Twitter, Facebook, and updates from check-in platforms like GetGlue. The multi-platform integration produced a visualization of all trending topics, keywords, and activities related to HBO on social media.[37] Vice president of social media Sabrina Caluori contrasted the activity on Connect with other online forums: "We know that there's an incredible amount of noise in the social media space today. More and more, users are looking for a strong POV. They're looking for curators. HBO Connect is a curated social media destination for fans of our brand and programming."[38] Caluori conflated the exclusivity of Connect with the prestige of HBO programs. In this construct, those with the best taste could escape the "aggregation" and "noise" of Twitter and participate in the "highly-curated" Connect experience. Of course, HBO had no problem contributing to the so-called noise on Twitter when it used the platform for publicity rather than genuine communication with users.

To further support this approach, HBO's irregular interactions with users often occurred on weekdays, separate from the heavier promotion for Sunday programming. On Thursday, April 24, HBO sent more than thirty tweets in rapid succession, all with the same text, to individual users who had praised early episodes of *Valley*: "@HBO: We're glad you like #SiliconValleyHBO! Please DM us your full name and address so we can send you some swag from the show." By sending these tweets directly to individual users, HBO ensured that they would not be easily visible in the main timeline unless followers sought out all of the network's activity. The tweets asked users to private DM (or direct message), signaling HBO wanted to keep the conversation away from public brand communication. Moreover, the identical nature of each tweet framed a standardized message as personalized engagement. Social platforms enable brands to generate this kind of false interaction en masse, undercutting any suggestions of sincere conversation. Perhaps users who received "swag" benefitted from their interaction with HBO's account, but the giveaways had value for the network as well. By offering fans free merchandise, HBO implicitly encouraged those fans to tweet about their "direct" contact with the network. Predictably, when this happened, HBO quickly retweeted fan gratitude, as it did earlier in the

HBO ✓ @HBO · Apr 24, 2014
Replying to @missriss0187
@missriss0187 **We're glad** you like #SiliconValleyHBO! Please DM us your full name and address so we can send you some swag from the show.

HBO ✓ @HBO · Apr 24, 2014
Replying to @hueknewit
@hueknewit **We're glad** you like #SiliconValleyHBO! Please DM us your full name and address so we can send you some swag from the show.

HBO ✓ @HBO · Apr 24, 2014
Replying to @MC_SimpsonFaxio
@MC_SimpsonFaxio **We're glad** you like #SiliconValleyHBO! Please DM us your full name and address so we can send you some swag from the show.

HBO ✓ @HBO · Apr 24, 2014
@HellsCannon **We're glad** you like #SiliconValleyHBO! Please DM us your full name and address so we can send you some swag from the show.

Sample of HBO's "engagement" with users on Twitter in 2014: an oft-repeated offer of free merchandise to *Silicon Valley* fans who had previously sent positive tweets about the series.

month on April 20, complete with a photo of the *Valley* T-shirt: "@lorddaveed: Thanks for the shirt, Richard. #SiliconValleyHBO @HBO." HBO's retweet positioned the network as both interactive and appreciative of fans, despite that a T-shirt giveaway is one of the most basic promotional strategies around. As Anderson notes, HBO cannot "afford to be an occasional-use medium" and that it "need[s] people on a regular basis" to sustain subscriptions.[39] This midweek pseudo-conversation gave people little reason to pay attention beyond Sunday nights. But that the engagement was inconsistent, standardized, and pushed to private channels only underlined that HBO did not merely, as Lury argues, "construct and manage" relationships with fans.[40] It also dictated and controlled the terms of the engagement to most benefit the network's existing reputation.

With the focus on new and original programming, HBO regularly ignored the content that made up the majority of its daily schedule: licensed feature films and reruns. HBO historically has been successful at "multiplexing," or repurposing, all of its offerings across its system of networks.[41] The repeated re-airing of new episodes has helped many HBO original series find an audience over the years. Likewise, the network still spends millions annually to acquire the home distribution rights for popular or acclaimed theatrical film releases. But while reruns and licensed films have a role in HBO's overall programming package, they do not typically fit within what Santo refers to as the network's "discourse of exclusivity."[42] Of course, HBO's tweets situated Sunday night as the prime destination and original programming as the main appeal of that destination. HBO even used weekdays to tweet about new episodes to come the next Sunday, thereby further marginalizing the licensed films and reruns prevalent on the daily schedule. The scant references to licensed films in the spring 2014 period embodied familiar branding logics, as HBO only tweeted about recently acquired films featuring Hollywood megastars:

> @HBO: "I like large parties, they're so intimate." Another chance to watch The Great Gatsby with @LeoDiCaprio starts now on #HBO2. (March 19)

> @HBO: Another chance to watch @paulfeig's The Heat starring Sandra Bullock and Melissa McCarthy starts now on #HBO. (March 30)

The choice to promote licensed films in the same fashion as original programming stressed the importance of star power and auteur figures to the HBO brand. Still, after the April 6 premieres of *Thrones*, *Veep*, *Valley*, and *Last*

Week Tonight, HBO sent no tweets promoting its film library. With vital originals on the air, HBO had little promotional bandwidth for licensed content. The disregard of licensed films reveals how HBO's economic imperative and brand image converge. Original programming has discursive value, situating the HBO brand around quality. But it also helps attract new subscribers each quarter and generates revenue in foreign and secondary distribution markets. The film library offers far less, both discursively and economically.

HBO's Twitter activity had a relatively narrow and conventional scope in spring 2014. Tweets were crafted with the "Not TV" brand in mind: original programming, Hollywood star power, and discourses of distinction, even in promotional blips or quick moments of pseudo-engagement with users. The above examples demonstrate what content and periods HBO purposefully pushed followers toward and what it avoided altogether. But HBO's evolving approach to streaming video—and its nascent competition with Netflix—inspired the company to also evolve its social branding strategy to a new type of performed authenticity and clearer embrace of Twitter vernacular.

NOW SOMETHING DIFFERENT

HBO has been viewed as the standard-bearer for quality television since the late 1990s, due in no smart part to its immense resources. As a critical cog in the WarnerMedia conglomerate, HBO has spent more—to attract the biggest stars, to build the most impressive sets, and to foster the most extensive branding campaigns. HBO's economic advantages are more pronounced considering the choice to schedule originals only a few nights a week, freeing the network to spend more on fewer projects and position those projects as significant events. As such, the streaming revolution initially did little to disrupt HBO's position. As Disney, News Corp, and NBCUniversal partnered to bring day-after episode streaming access to Hulu, and Netflix and Amazon added thousands of hours of television to their libraries, HBO kept its programming behind authenticated paywalls. When HBO unveiled HBO Go, an on-demand streaming portal for existing subscribers, in 2010, industry insiders speculated if it would lead to an à la carte version.[43] But in the years after Go's introduction and immediate success, HBO and Time Warner executives remained steadfast in that HBO should remain only as a premium service within the cable and satellite infrastructure. As late as May 2013, Time Warner CEO Jeff Bewkes claimed that the streaming video marketplace "was not sufficiently big enough" for HBO to explore à la carte options to compete with Netflix and the rest.[44] For a time,

HBO's bet on its long-standing model paid off. The HBO Go app was downloaded more than three million times in its first few months of release.[45] The press regularly covered Go-related ephemera like password sharing and the platform's crashing due to overuse during live airings of *Thrones* and *True Detective*.[46] As increased time- and place-shifting destabilized live viewing and cable subscriptions, HBO used Go to further eventize its series and Sunday nights.

Amid HBO Go's success, Netflix began to build a stable of original series. In 2012, Netflix famously outbid HBO for the first-run rights to *House of Cards*, but that same year it announced a new season of cult Fox sitcom *Arrested Development* and *Orange Is the New Black*, a new project from *Weeds* creator Jenji Kohan.[47] At the time, Netflix's Ted Sarandos explained that the fastest way to topple HBO would be to outspend it on original programming and attract top Hollywood talent. Early partners reinforced this sentiment. Kohan called her work with Netflix "even more liberating" than working for a premium cable outfit like Showtime.[48] Along with mimicking HBO's spending and branding, Sarandos championed Netflix's all-at-once distribution as the key to winning the new competition. Situating the distribution model as a way to match existing viewer habits, Sarandos claimed, "The move away from appointment television is enormous. So why are you going to drag people back to something they're abandoning in huge numbers?"[49] The critique of appointment television was another shot at linear television networks like HBO, which built its industry supremacy by prioritizing a single night of prestige programming on Sundays.

Netflix's big spending and novel distribution tactics produced significant attention from the press, especially in light of increased anxieties about cord-cutting. Pay television services lost nearly one million subscribers in the first half of 2015, and by the end of that year, one-fifth of all US households were without cable or satellite altogether.[50] Netflix did little to quell the growing industry narrative regarding its competition with HBO, even long after it had surpassed HBO in paying US subscribers, going as far as repeatedly referencing the competition in "long-term view" statements to investors.[51] After pushing back for years, including once suggesting that Netflix posed no threat and later arguing that the two companies offered "complementary" services, Time Warner CEO Bewkes finally relented.[52] He admitted that "tech companies" had embraced streaming "faster and better" than HBO.[53] With the rollout of HBO Now in 2015, HBO finally made its programming available to cord-cutters/nevers. It also embraced platform authenticity, blending its renowned prestige brand image with a newfound commitment to viral hashtags and spreadable memes.

The most noticeable change came in how 2017 HBO used GIFs and videos in its tweets. To a degree, HBO's progression in this regard embodied broader improvements in Twitter's ability to handle the formats without having to link to an external source. But it is also fair to assert that both Twitter and HBO embraced GIFs because it had become the leading way to respond to everything—live events, popular media, daily life—online. The GIF, which presents a brief, looping, and silent moving image, has a long history but emerged as a premier component of conversation on Tumblr, as well as the appropriation into "listicles" produced by digital news companies like BuzzFeed. As Jason Eppink writes in his history of the form, by 2011, GIFs were being posted "in response to, and often in lieu of, text online."[54] GIFs capture moments that express universal emotions, hyper-specific experiences, and convoluted metaphors, leading Michael Z. Newman to call them "distillations of pure affect."[55] Often paired with relevant brief text, GIFs now power so much of the chatter and relationship-building on social platforms like Twitter. Eppink claims that the GIF is commonly used in two types of posts. The first type is "Actual" reaction GIFs, which respond to real events or other users' posts. The second type is "Hypothetical" reaction GIFs, which propose a theoretical scenario that the GIF illustrates or comments on, perhaps best personified by the "HIFW (How I Feel When)" or "MRW (My Reaction When)" themes.[56]

For a brand like HBO to communicate through GIFs, it must perform the personal-but-universal experience that defines the HIFW or MRW structures. Yet, the use of "I" and "my" challenges any brand, let alone one that has presented itself as a prestigious monolith. HBO reformulated the familiar structures, orienting them toward the audience, and with programming in mind.

@HBO: When someone tells you they don't watch your favorite show. #HBO (May 12)

@HBO: When you just finished your favorite show. #StartSomethingNew #HBO (May 16)

@HBO: When someone says they've never seen #TheWire. #HBO #TheWire15 Watch the first episode for free: http://bit.ly/WatchTheWireFree (June 2)

@HBO: When you're laughing so hard at @SiliconHBO and @VeepHBO, you forget your sad because the seasons are ending. #HBO (June 25)

Here, the "When . . ." formula hinged on watching, or an appreciation for, HBO series. Tweets were accompanied by a GIF from the relevant HBO series, recontextualized to punctuate the meaning of the text. The first featured a disgusted Richard (Thomas Middleditch) from *Valley*, judging the theoretical person's taste in television. The second displayed *Big Little Lies*' Madeline (Reese Witherspoon) excitedly screaming, "I want more!" regarding HBO content. The third highlighted a much-used *Wire* image, with Wee-Bey (Hassan Johnson) reacting in horror to the person's lack of familiarity with the acclaimed cult classic. The final tweet highlighted *Valley*'s cast sharing festive drinks in honor of season-ending episodes for the series and Sunday night partner *Veep*. Eppink argues that an appeal of the GIF is the lack of authorship as people share and recontextualize the meaning of an image with each post.[57] But the authorship was clear; HBO "made" these images on television and repurposed them to speak to an affective experience filtered through the prism of the HBO brand. The tweets denoted that HBO was *so central* to the lives of the imagined user that they would be, in sequence, 1) repulsed by a friend who does not love the same series, 2) equally shocked to learn that friend has not seen *The Wire*, 3) insatiably hungry for more episodes of all HBO series, and 4) devastated that *Valley* and *Veep* were ending for the summer.

HBO's GIF use dovetailed with its increased direct engagement with users. Whereas 2014 HBO tried to guide the conversation to DM or Connect, in 2017, it willingly and publicly responded, often using GIFs to punctuate a point. When user @lslrn noted their enjoyment of an episode of *Veep* (using the #Veep hashtag), HBO quickly responded with "IT. GETS. EVEN. BETTER.," as well as a GIF from the series where a character exclaims, "Yeah, I gotta call you back. Something amazing is happening." Likewise, when user @cinema416 tweeted, "My favorite part of *The Sopranos* is those instances between all the yelling & fighting where the cast says a lot with no words. @HBO," the network responded simply with a reaction GIF of Paulie Walnuts (Tony Sirico) silently raising his eyes toward someone out of frame, as if to say, "We see you." In these examples, HBO shared brief and informal exchanges with users similar to how any two people would talk online. Yet, HBO's contribution to the conversation only appropriated the vernacular of Twitter to spread its content further onto the timeline.

BINGEING AND STREAMING COME TO HBO

Television's migration to streaming portals and the growth of binge-watching aided the evolution of HBO's social brand. Despite the success of the subscriber portal HBO Go, HBO mentioned any combination of "stream" or "binge" just once in the spring 2014 period and only ten times across all of 2014, and instead relied on the more conventional "watch." By 2017, however, HBO had embraced the modern verbiage, signaling an awareness of how people watch—and talk about watching. HBO sent twenty-six tweets referencing stream or binge in spring 2017, and nearly forty during the first half of the year. HBO addressed stream or binge in an array of contexts, from frequent updates about just-aired episodes to more spirited examples free of a specific broadcast or timeslot. On April 30, HBO promoted its Sunday night lineup of *The Leftovers*, *Valley*, *Veep*, and *Last Week Tonight* not with a big star or auteur but instead an image of a "talking jar": "@HBO: Friendly reminder: streaming starts promptly at 9:00, shoes off at the door, no talking. #HBO." In tweets like this, HBO negotiated its old and new brand identities. The tweet arrived promptly at 9 p.m. EST as a familiar real-time notice of immediately upcoming live airings. But while the tweet centered live Sunday programming, it also assumed that many viewers would be streaming via HBO Go or Now; controlling access to how people watched was no longer critical. The tweet's mode of address also simultaneously embraced and tried to suppress the communal experience shared by multi-screen viewers. The tweet and image publicly and purposefully shamed talking during HBO programs on Twitter, a platform known for its live chatter, thereby hoping users would pick up on the snippet of performed irony.

Tweets in this realm also accepted that people were streaming beyond the live airing. In a few instances, HBO stressed the streaming or bingeing of projects that had already ended their respective runs on the schedule like mini-series *Mildred Pierce* and *The Night Of*:

> @HBO: It's the perfect time to fall back in love with Mildred Pierce. Streaming the #HBO mini-series until further notice. (April 8)

> @HBO: One episode will change the way you think about . . . well, everything. Let the bingeing begin. #TheNightOf (June 22)

The first tweet's use of "fall back in love" pointed to the process of rewatching *Mildred Pierce*, while the second expressed the practice of watching *The Night Of* for the first time. In both cases, HBO showed an awareness of the new normal where viewers stream "until further notice" and binge to catch up on long-completed projects. HBO directed viewers not to the linear television schedule but rather to the streaming library of always available quality content. Many tweets of this type also used humor to trivialize the efforts of streaming and bingeing:

@HBO: Gave up bingeing for lent? We'll just leave this here. #HBO #TheYoungPope (April 14)

@HBO: Friends don't let friends brunch without a binge recommendation. #HBO (May 6)

@HBO: Must sleep . . . but . . . can't . . . stop . . . watching . . . #HBO #CantStopWatching (May 17)

@HBO: Do you define your #HBO binge in episodes . . . or snacks? (June 24)

The initial two tweets included GIFs from *The Young Pope* and *Big Little Lies* intended to punch up the humor of the text. In the first, the titular pope (Jude Law) is furiously juggling, which, combined with the words, proposed that committing to a no-binge Lent was a risky proposition requiring determination. The notion that HBO wanted to "just leave this [the GIF] here" was meant to be a form of social media temptation. The next tweet showed a GIF of Witherspoon's Madeline emphatically telling a friend, "You're going to love it." The text and image, freed from their diegetic context, worked together to conjure a familiar experience of one friend evangelizing to another about their latest media binge. The insinuation was that bingeing is a centerpiece of many brunch conversations, and that contemporary friendship is, in some part, defined by making appropriate streaming recommendations.

The third and fourth tweets spoke directly to the habits of streaming and bingeing. Containing a GIF of an exhausted couple on the couch, the "Must sleep" tweet tried to represent what it *feels* like to be in the midst of a binge session. The GIF displayed the couple in a medium-wide shot, with one woman sprawled on the couch, asleep, and eyes covered. A close-up of the second woman showed the physical effects of bingeing, as she intently stared at the glow of the screen and refused to blink in fear of also falling asleep. This imagery was presented with a fuzzy shadow-like effect and blue and yellow hues, conjuring the trance-like sense of all-night viewing. The fourth tweet, meanwhile, revealed a photograph of snacks, Chinese takeout, empty Red Bull cans, *Thrones* champagne, and a television remote. Unlike the previous examples, this tweet united culture's two visions of bingeing: that related to eating and to watching. But instead of stigmatizing these behaviors, HBO situated bingeing as a badge of honor—as long as HBO programming was involved. Together, these tweets drolly celebrated modern viewer habits without pushing too hard about exactly what to watch.

Still, HBO's celebration of streaming and bingeing materialized leading into, or during, the weekend. While consumers stream and binge at all points of the week, HBO hoped that its followers would spend the weekend streaming its library content.

@HBO: What a long weekend of bingeing @GameofThrones feels like. (May 28)

@HBO: Sunday's loop: sleep, snack, stream, sleep. #HBO #Westworld (June 25)

Both the *Thrones* and *Westworld* tweets were bolstered by GIFs to advance the idea that HBO recommended these practices. The former featured Daenerys triumphantly riding one of her dragons as it soared into the sky, gesturing that a long weekend binge was an exhilarating experience. Meanwhile, the latter showed the repetitive behavior of the robot Dolores (Evan Rachel Wood), parsing a successful Sunday down to the essentials of sleeping, snacking, and streaming. Here, HBO used humor to place its series into the weekend routine. Writing of HBO's early web campaigns, Johnson argues that the network "aim[ed] to provide a link to HBO content where the viewer is most likely to find it."[58] Modern HBO, meanwhile, wants promotions to match the tone and style of the discussions already circulating about the brand. If, as Polan has argued, HBO tries to reach the urban professional who "needs not only meaningfulness and substance but also hipness, newness, and cutting-edge innovation," then the embrace of streaming and bingeing embodied newer forms of hipness and innovation.[59] For HBO, the tension emerged in retaining the sense of exclusivity while also conceding broader cultural changes. Tussey contends that ephemeral snack content also "keeps the audience invested in more lucrative traditional media texts."[60] This updated vision of the brand was still centered on Sunday, the weekend, and quality—markers intended to drive folks to HBO programming. But assumptions about how people access or consume that programming had changed.

Johnson argues that the web enables networks to distribute content into new platforms and extend the brand to new arenas.[61] However, less is said about how brand extensions require new strategies or new ways to refer to standard practices. HBO's celebration of both streaming and bingeing—as protocols and as verbiage—displayed the continued influence of what people do with brands; the broader acceptance of streaming and bingeing *necessitated* that HBO would also accept those methods of viewership. This evolution illustrates that successful brands like HBO agilely morph and position themselves around changes in the everyday consumer experience.

CELEBRATING AND SURVIVING WITH HBO

HBO also demonstrated a newfound interest in major and minor holidays, particularly those that could be tied in some way to its original programming. This ceremonial content took a comedic tone geared toward easy, hashtaggable sharing.

> @HBO: Let's roll. 🚴 #HBO #NationalBikeMonth #HighMaintenance (May 4)

> @HBO: Celebrating #CincodeMayo at the HBO office today! 🎉 #HBO #TheSopranos @Veep @GameofThrones (May 5)

> @HBO: Some donuts fresh from the 🍩 oven. Anything calling your name? #NationalDonutDay #HBO (June 2)

> @HBO: Celebrating #BestFriendsDay with some Ballers #HBO @ballershbo (June 8)

Each tweet included a GIF, photo, and multiple hashtags—three also featured a festive emoji—to position HBO as a hip brand participating in the fun online celebrations. The tweets mentioning *High Maintenance* and *Ballers* offered decontextualized GIFs of the comedies' main characters on bikes and horsing around respectively, reconstituting the images' meaning within the frame of the holiday. The Cinco de Mayo and National Donut Day tweets, meanwhile, displayed original artwork to promote HBO programs more explicitly. The former showed an image of three hands near a margarita machine, each representing *Sopranos*, *Thrones*, and *Veep*. The latter showcased six donuts, five honoring a series (*True Blood*, *Veep*, *Thrones*, *Westworld*, *Sex and the City*) and one featuring HBO's logo. The Cinco de Mayo tweet's reference to "the HBO office" proposed that the HBO universe was an *actual place* where fictional characters from popular series not only commingled but also participated in stereotypical "work party" events. Though the National

Donut Day tweet was less direct, it still visualized an office setting with the donuts on display in a conference room table basked in the glow of fluorescent lighting. Lynn Spigel and Max Dawson argue that the "'social arrhythmia' of the new 24/7/365 post-industrial economy" incentivizes television networks to eschew conventional daypart strategies and explore more "flexible leisure" options, including during the workday.[62] But what is particularly fascinating about these tweets is that they exist not only as flexible, snackable content for workplace distraction but also that they are pointedly *about* the workplace. They perform a kind of meta-commentary on the banality of postindustrial office culture and the desperation of procrastination behaviors. And as a branding device, the tweets posited that HBO understood the quotidian experience of the workplace and showed how easily HBO fit into that setting and its related routines.

HBO's self-reflexive Cinco de Mayo promotional tweet in 2017 featuring the disembodied hands of Tony Soprano, Jaime Lannister, and Selina Meyer.

If weekends were for bingeing in HBO's world, then weekdays were often dedicated to surviving, whether by cliché workplace celebrations, or simple perseverance, to make it to the next weekend streaming session. In these tweets, HBO was an empathetic partner helping followers get by with brief moments of camaraderie:

@HBO: Let's get through this week together. #HBO #MondayMotivation (May 8)

@HBO: Friday. Because no one ever exclaimed, 'Thank goodness it's Wednesday!' #HBO (May 19)

@HBO: Clawing our way to Friday. #HBO #BigLittleLies #FYC (June 8)

@HBO: When you haven't had your Monday morning coffee yet. #HBO #Mondays @SiliconHBO (June 19)

Though HBO used the #MondayMotivation hashtag to appear in searches, its "motivation" was intended to be tongue-in-cheek. The "Let's get through this week together" tweet featured a GIF from *Veep* with former president Selina Meyer (Julia Louis-Dreyfus) and beleaguered assistant Gary (Tony Hale) sharing an awkward moment of agreement. The coffee tweet tapped into the cultural meme of needing coffee to start the day/week. Still, it partnered the text with a GIF of *Valley*'s antagonist Gavin Belson (Matt Ross) as he dejectedly destroyed expensive art, thereby comedically escalating the meme in an HBO context. Other tweets in this category addressed stereotypes related to their respective weekdays: Wednesday is a day that few enjoy. Friday is a day so desirable that people are willing to "claw" their way to it. HBO posted each of these tweets in the morning or midday. The posts tried to motivate followers into surviving work on Monday and Wednesday and rejoiced alongside those celebrating the end of the workweek coming on Friday. Grainge argues that "ambient communication [is] a means of achieving intimacy ... inserting brands into an expanded range of everyday spaces."[63] Despite their corporate origins, social platforms, and the devices and contexts in which people access them, are undoubtedly everyday modern spaces. HBO's focus on workweek survival and weekend celebrations erected an intimate relationship with followers based less on programs—which were still present in GIFs and images—and more on the recognition of, and empathy for, an imagined "typical" labor experience.

HBO's Twitter focus on the everyday experience effectively branded its most essential day, Sunday, as *even more* essential:

@HBO: Sunday is and always will be our special day. #HBODate (May 8)

@HBO: Watching HBO on Sunday night is the difference between being in on the jokes and being on the outside looking in. (May 28)

@HBO: Getting ready with the wine and the watchlist ... because it's Sunday night. #HBO (June 4)

Here, HBO delivered the familiar Sunday messaging but in a more conversational manner. The first tweet, featuring another awkward moment with *Veep*'s Selina declaring, "This is my house," emphasized the historical *and* future significance of the network's Sunday night programming. The second tweet indirectly cited HBO's long-running contribution to workplace watercooler chatter. The tweet punctuated the claim with a comedic GIF of *Valley*'s resident sack Jared (Zach Woods) looking longingly out the window—in this context, at all the great television he missed on Sunday. The final post barely bothered with any explanation and instead relied on the common HBO refrain of "because it's Sunday night." Like the "It's Not TV" slogan, the tweet played on an assumed collective cachet among subscribers: HBO Sundays remain in a category of their own, and sharp viewers already know what to do. This was far from a new ploy; as Santo asserts, past HBO promos have "intended to create audience identification with Sunday night as *belonging* to HBO.... ensur[ing] that new series both receive instant viewers and cultural cache when slotted into the Sunday night schedule."[64] These tweets notably did not celebrate a particular series or even promote specific soon-to-air episodes. They situated Sunday as the centerpiece event, suggesting that *all* HBO Sundays deserved special attention. The self-referential and self-reflexive humor of the GIFs only accentuated the in-group status of those viewers who both got the jokes and understood the significance of HBO Sundays.

The eventizing of the minor holidays and the everyday experience became more expressly promotional when HBO capitalized on the first day of summer with an ironic synergy for *Thrones*' seventh season and #WinterIsHere. Plugging the hashtag on June 21 to commemorate the release of a new trailer for the fantasy serial's anticipated next season, HBO tweeted:

@HBO: It may be the first day of Summer, but #WinterIsHere on July 16. Watch the new @GameofThrones Season 7 trailer.

HBO crafted more profound synergistic unity by having winter "take over" other series, as the account published familiar images from *Sex and the City* and *The Wire* overlaid with the patented blue-gray winter color scheme of *Thrones*. The *Sex and the City* tweet showed the four lead characters with cold air emerging from their mouths marking the temperature drop and the accompanying line, "New York is about to experience a cold snap." Hashtags—#WinterIsHere, #GoTS7, #SATC, and #HBO—appeared on the image to indicate that winter and *Thrones* had thoroughly swept through the HBO universe and established its supremacy. To celebrate this faux holiday, HBO encouraged a small bit of user interactivity into the campaign as well: "@HBO: Psst @GameofThrones fans, unlock a winter surprise from your favorite HBO shows. Hint: Get creative with your emojis. #HBO #WinterIsHere." HBO prompted users to "unlock" content with a "hint," turning the posting of emojis and the hashtag into a vague game that helped the #WinterIsHere trend. The resulting "surprises" were images like the above *Sex and the City* graphic and new character posters for the upcoming season of *Thrones*. The call for specific emojis represented a small increase in engagement to better align with other brands on Twitter. Still, HBO constructed a scenario for fans to participate that mainly only produced more promotional content. Carah claims that "brands depend less on consumers as participants in the creation of specific meanings and more on their capacity to act in ways that can be watched and responded to."[65] I would argue that brands on Twitter facilitate not only the creation of specific meanings but also the creation of specific outcomes. #WinterIsHere illustrated how hashtags could generate these outcomes, particularly when they are attention-oriented. While fans participated in the "unveiling" of new *Thrones* material, they ultimately helped HBO secure a burst of digital attention. Carah also posits that the migration of brands to social platforms has diminished their ability to help consumers create "an authentic reflection of their identities and ways of life."[66] Instead, he argues, brands connect "material cultural spaces, mobile devices, and participatory data-driven media systems" to "modulate relationships of attention in an open-ended way."[67] But while social platforms escort users toward controlled forms of participation, brands pretend to circumvent those controls in the form of a performed authenticity. Indeed, brands push even harder to present as authentic and to build relationships because of the visible restrictions embedded in social platforms—*this particular interaction* with emojis or hashtags is more meaningful than the artificial examples elsewhere online. Authenticity is, therefore, translated through discursive style and technocratic symbols.

HBO AND PRIDE

No campaign represented HBO's modern blend of holiday celebrations, platform authenticity, and conventional publicity like its celebration of June's LGBTQ+ Pride Month. Predictably, HBO changed its profile photo across social media from the familiar white logo on a black background to an LGBTQ+-friendly rainbow logo. The network expanded its digital activism with a month-long promotional blitz, including a tagline, "It's what connects us," a group of rainbow-filtered clips from relevant HBO series, and supportive messages for Pride and LGBTQ+ equality. Of course, HBO stamped every piece of social content with the requisite #Pride and #HBO hashtags. HBO sent more than a dozen tweets in this format in June, including:

> @HBO: When someone "gets you." Understanding: It's what connects us. #HBO #Pride #SixFeetUnder" (June 3)

> @HBO: Sometimes, you just know. Love: It's what connects us. #HBO #Pride #BessieHBO @IAMQUEENLATIFAH" (June 4)

> @HBO: Gender identity is who you are. Sexual orientation is who you love. Understanding: It's what connects us. #HBO #Pride @LastWeekTonight (June 15)

> @HBO: "The world is changing. I get to be whoever I want." Authenticity: It's what connects us. #Pride #HBO" (June 26)

Brian L. Ott argues that with series like *The Wire*, HBO "[made] entertainment political and politics entertaining."[68] But here, HBO tried to go beyond showing "entertaining" politics and take a stance about Pride as a valuable political event and LGBTQ+ equality as a worthwhile sociopolitical cause. In the first two tweets, the messages of love and understanding were boosted by GIFs of LGBTQ+ characters in various states of affection. The other tweets addressed equality a bit more overtly with clips of non-celebrities discussing evolving ideas about gender identity. Ron Becker argues that HBO has used diverse representation to appeal to politically progressive viewers, but also to build long-term loyalty with educated, high-income LGBTQ+ audiences.[69] The Pride campaign illustrated the uncertainty surrounding the public embrace of politics by brands. On the one hand, HBO tried to raise legitimate awareness for Pride and the LGBTQ+ community. On the other hand, how HBO did this activism—through slogans, hashtags, and links to programs—was as much about reaffirming the progressiveness and cultural cachet of HBO as it was for making important points about Pride or equality. Indeed, the use of "connects" functioned as another prompt for Twitter users to connect with HBO, its series, and its streaming portals as much as it asked them to connect with one another.

HBO tried to make its activism actionable by attending the New York City Pride parade:

@HBO: The @HBO family is all ready to support and march alongside @amFAR at the NYC Pride Parade. #Pride2017 (June 25)

The photo showed the HBO "family": a multi-cultural group of six smiling millennial-age people, all donning a shirt for amFAR, an international foundation supporting AIDS research. Though it was unclear who represents what company, everyone in the photo stood on a bustling city street with active parade participants behind them. The tweet personified the idea of HBO as an actual place, but instead of a place where fictional characters celebrate Cinco de Mayo, it was one where real people worked for political change. HBO's "participation" and post also created a showcase for Pride, amFAR, AIDS research, and LGBTQ+ issues. But, again, HBO also intended the showcase for those organizations to create a positive halo effect around its brand (what has increasingly been called "rainbow washing").[70]

The Pride campaign embodied Banet-Weiser and Roopali Mukherjee's "commodity activism," where people are forced to become "consumer-citizens" to practice moral virtue and civic engagement through modes of consumption.[71] As Laurie Ouellette argues, corporations drive this commodified activism, promoting "corporate social responsibility" internally and encouraging citizens to "do good" by wearing the "right" products or donating to the "right" causes.[72] Ouellette surveys corporate branding guides that encourage companies to present a "real dedication to being part of human solutions around the world" because increased interest in activism "can be harnessed as a form of market intelligence."[73] That corporations perform the responsibility of turning customers into better people is a condition of a culture where consumption, or tacit approval of corporate social responsibility, functions as political engagement. Ouellette notes that whether these companies believe they can convince people of their morality is beyond the point. Instead, the goal of campaigns like HBO's Pride initiative is to get people to "perform one's duty as a citizen" within the brand context to produce "an ethical surplus that can be recuperated as brand value."[74] Once corporations make moral considerations—each with slogans, hashtags, and

GIFs—the ultimate political potency of the issue remains muted; meanwhile, the companies still benefit from their performance as what Ouellette calls a "citizen brand," or in popular discourse is known as performative allyship.

For HBO, this ethical surplus is very relevant. The network's economic model *necessitates* that people continually feel positively enough about the brand—whether via the programming, social justice initiatives, or something else entirely—to keep subscribing. By sharing its actual participation in a significant Pride event on social platforms, then, HBO positioned itself as a company with *real people* and *real values* that align with the progressiveness of its programming. Significantly, it also proposed that HBO was not just a social media "slacktivist" raising awareness with a temporary rainbow profile photo.[75] Finally, this strategy also underlined the all-encompassing totality of contemporary branding strategy. Promotional material does not just fully occupy digital or physical spaces separately but rather flows between those spaces with ease. Indeed, *every* space is an opportunity to build or reaffirm the authentic voice of a brand and the relationship between brand and consumer.

"ALWAYS BE THERE": CONCLUSIONS ON HBO

Two final tweets are worth mentioning regarding how HBO shifted its social practices:

> @HBO: The episodes you missed are always there when you need them most. #HBO (April 24)

> @HBO: Friends come and go, but your favorite episode will always be there. #HBO (May 9)

Like many of the tweets described in this chapter, these examples operate on multiple fronts. "Catch up later" and "always be there" naturally nodded to the time-shifting habits of the audience and the always-available libraries of HBO Go and HBO Now. But in stating that it would *always* be there for subscribers, HBO emphasized the affective relationship between the network, its programs, and its fans. The first tweet intensified the connection with a video featuring a fan who received flowers and a card reading, "Missed you last night. Let's catch up later. Love, HBO." The second presented a *Westworld* GIF where one character grabbed the hand of the other, stressing the metaphorical bond between HBO and its audience. In both cases, there was no subtext. HBO pitched its brand and its content as an improvement over the messy complexities and disappointments of human relationships. The network promised to provide beloved episodes and series, and an "affectively significant relationship," always.[76]

At first, HBO experienced some challenges as its established brand image evolved on Twitter—chiefly striking a balance between conventional prestige promotion and more modern social engagement with fans and followers. For years HBO's immense financial resources and broader cultural cachet continuously supported the network's premium subscription business model. Branding played a crucial role in HBO's success. The "Not TV" campaign remains perhaps the television industry's most notable promotional initiative. But as more competitors entered the marketplace with an appropriation of the "Not TV" playbook of big spending and quality TV discourses, HBO was forced to consider alternative methods to reach—and relate—to the next generation of potential subscribers. HBO turned to Twitter, which was already playing a vital role in the Social TV realm, to facilitate new forms of attention for its programming and brand. Instead of exploiting Twitter as yet another place to dump traditional promotional content, HBO eventually found notable ways to participate in the evolution of Twitter and, by proxy, develop a more visible relationship with its followers. A big piece of this affective relationship-building involved combining what I would call *default fan voice* with *default Twitter voice*. Every chapter in this book has underlined corporate attempts at constructing a contained and reactive fan with one Social TV strategy after another. HBO's more modern Twitter profile exhibits an actual effort to fit television content into the contours of a social platform. Indeed, the diversity of tones (from detached irony to sincere activism) and subject matter (from binge-watches to office holiday parties) speaks to the flexibility and polysemy required of companies that seek the spotlight on Twitter. Thus, even though the new approach did not entirely erase HBO's historical prestige image based on Sunday nights, big-time stars, and cultural capital, it did evolve HBO's brand from a premium cable network to a premium content provider.

As part of HBO's relational approach, the performance of platform authenticity acknowledged the outside influences on the brand: consumer behavior, industry trends, and changes in online vernacular. Banet-Weiser argues that brands invent Raymond Williams's "structure of feeling," and I would argue that this phenomenon is fortified over time.[77] Media brands are particularly well-positioned to develop this bond with consumers because of the countless memories they share. People remember their favorite moments from favorite episodes, when they cried over a beloved character's death in a film sequel, or, increasingly, when a company takes a political stand in the name of social justice. For HBO, television—with its cliffhangers, significant deaths, and romantic detours—established the structure of feeling. HBO's performance of social genuineness, then, served to remind fans of these cherished or heartbreaking or tolerant moments through a barrage of GIFs, hashtags, clips, references, and in-jokes. The network's platform authenticity tried to *deepen* the relationship, notably between periods of engagement with television.

The multitude of content options brought forth by technological change, and industry trends have upended what Roger Silverstone calls television's "veritable dailiness" and chipped away at the permanence of networks as organizational structures.[78] The conventional viewing rituals and frameworks that helped television integrate into everyday life matter less than ever. For much of the Social TV era, the television industry viewed social platforms as a tool to re-instill the value of those rituals and frameworks, particularly regarding live viewership. Although some of those Social TV initiatives found success over short periods or with niche audiences, the broader shifts in audience behavior and content distribution could not be slowed by multi-platform, hashtag-centric campaigns promising potential engagement with celebrities.

But as Paddy Scannell argues, networks have *always* had an "unenforceable" relationship with audiences.[79] In that regard, new technologies serve as tools for the most opportunistic networks to sustain relationships with the audience in new ways. For HBO, that means using Twitter to illustrate an understanding of how rituals involving television have evolved—including not only how people watch it but also how they talk about it—while keeping itself at the center of those evolutions. It also means circulating a repertoire of social content to advance the brand beyond permanence toward continual ephemerality, or omnipresence, and mainly sustaining bonds with audiences by *being everywhere all the time*. Scannell argues that broadcasting uses "ordinary, everyday, mundane conversation, or talk" to construct a collective imagination of television as an "all-day-everyday" experience.[80] The modern version of this is a multi-platform performance of omnipresent authenticity, where programming, social updates, and corporate activism work in concert to encircle consumers in a never-ending ecosystem of content and, as Tussey claims, to encourage them to "spend 'snack time' with [corporate] intellectual property."[81]

HBO's emergent Twitter practices display how, in a Social TV landscape where so many forms of content were always-already available, television—or television-related content—can still, as Henri Lefebvre argues, "occupy" and "personify" everyday life.[82] In Lefebvre's view, television pushes hard to create essential moments of the "repetitive monotony" in the everyday, orienting people away from meaningful experiences outside of consumption.[83] With exclusive and searchable hashtags, contrived holidays, and corporate social responsibility, companies like HBO simultaneously perform a genuine understanding of the everyday experience and present that experience as an event to be celebrated, in collaboration with our favorite brands. The everyday is, therefore, eventized—not unlike HBO's Sunday night programming—as long as it is filtered through the prism of a network brand identity. This brand-creep into the everyday aids companies like HBO in navigating the choppy waters of social media but also underlines an ambivalence about how corporations infiltrate and structure even the smallest moments of modern life.

Conclusion

EVERYDAY EPHEMERAL CONTENT

In the summer of 2017, Facebook and Twitter, two of the world's most influential social platforms, announced new expansions into original video content. Borrowing from the television industry practice of the upfront, where networks trot out new programs and celebrities to woo sponsors into buying advertising time for the following season, Twitter held a "Digital Content NewFronts" presentation in May 2017 for a litany of prospective corporate partners. There, Twitter revealed a partnership with Bloomberg Media to launch a twenty-four-hour news network known as TicToc for "an intelligent audience around the globe."[1] TicToc (renamed Bloomberg QuickTake in 2019 to avoid confusion with another emerging video platform, TikTok) would "broadcast" via short-form videos embedded into each tweet. Along with the around-the-clock news effort, Twitter announced "live-video" broadcasting rights to Major League Baseball and Women's National Basketball Association games and talk shows from Vox Media and *BuzzFeed News*. During the NewFronts event, Twitter CEO Jack Dorsey declared a simple goal for the new video rollout: "We want to be the first place anyone hears of anything going on that matters to them. That's our focus."[2] Leslie Berland, Twitter's chief marketing officer, explained to potential sponsors why a social platform known for chaotic conversations and viral hashtags would serve as a great place to broadcast live video content. As Berland said, "Twitter is about what's happening, and what people are talking about right now."[3]

Not to be outdone, Facebook introduced a new video initiative of its own in August 2017. Mark Zuckerberg's social media giant revealed Facebook Watch, a newly redesigned and centralized hub for video content. The press release announcing Watch described the functionality of the video portal in meticulous detail:

> We're introducing Watch, a new platform for shows on Facebook. Watch will be available on mobile, on desktop and laptop, and in our TV apps. Shows are made up of episodes—live or recorded—and

follow a theme or storyline. To help you keep up with the shows you follow, Watch has a Watchlist so you never miss out on the latest episodes.... We've learned from Facebook Live that people's comments and reactions to a video are often as much a part of the experience as the video itself. So when you watch a show, you can see comments and connect with friends and other viewers while watching, or participate in a dedicated Facebook Group for the show.[4]

Facebook's News Feed long had been riddled with short user-generated videos. But the company's plans for Watch included expanded production budgets for original long-form video series to be financed and produced in-house. Facebook delineated the two types of video content with awkward internal labels: "spotlight" for short-form videos and "hero" for 20–30-minute scripted and unscripted series.[5] Much like Twitter, Facebook signed deals with publishing and news companies like Condé Nast and BuzzFeed. It also, however, announced "hero" series like World Wrestling Entertainment's *Mixed Match Challenge*, *SKAM Austin*, an American adaptation of the popular Norwegian teen drama *Skam*, and *Ball in the Family*, a documentary series about the infamous Ball basketball clan.[6]

Twitter and Facebook's 2017 announcements were the latest in a long line of plans to integrate video content into their platforms. In 2016, Twitter paid $10 million for the live streaming rights to the National Football League's *Thursday Night Football*.[7] It previously helped facilitate the live broadcasting of political debates, election coverage, and award show red-carpet specials.[8] Facebook, meanwhile, announced an original "new shift to video" in 2015.[9] The company's newfound interest in hosting video content catalyzed sweeping changes in the digital news industry, which relies heavily on traffic from the Facebook platform. The news industry's resulting "pivot to video" saw numerous companies (including legacy publishers like *Vanity Fair* and *Sports Illustrated*) prioritize cheap short-form video content over traditional reporting, and, ultimately, see minimal returns when Facebook decided to reconfigure the platform's video-oriented algorithms.[10] Facebook also experimented with video content via Facebook Live, a feature that allowed anyone to create and distribute live content. Facebook Live became a go-to tool for live simulcasts of significant pop culture events and breaking news in local markets.[11]

Amid the didactic description of Facebook Watch's features and the lofty ambition of Twitter's broadcasting partnerships, the new initiatives notably did not reference one word: television. Still, the companies' video proposals served as yet another attempted remediation of television. Twitter and Facebook offered up a transformation significant enough to the point of

Facebook Watch promotional photo with the streaming portal's interface accessed on a television screen.

erasing television from their promotional discourse. Of course, the erasure of the word "television" could not hide that Twitter and Facebook partnered with talent from the television industry to build their video products or that the products relied on several television-related concepts. The remediation of television manifested in how Twitter and Facebook labeled their content (shows, series, episodes, broadcasts), their technological processes (network, watchlist), and their temporal focus (first place, right now, live). As the companies pitched their video endeavors as intrinsically social, they also spotlighted the shows and live broadcasts that would generate that conversation across timelines and news feeds. In fact, Twitter and Facebook appropriated the utopian discourses of prior remediation attempts. They situated their video initiatives as chiefly fan- and community-focused, with everyone slated to watch, chat, and share memes collectively in one space. Their promises inflated the disruptive potential of their video products to the point of parody. In the hands of Silicon Valley leaders like Twitter and Facebook, concepts with near-universal understanding like "watch," "shows," and "live video" get capitalized or stylized for artificial emphasis and import.

Twitter and Facebook's flirtations with video content also repeated the long-standing media industry strategy regarding original programming. In the 1980s and 1990s, cable networks pivoted from licensing television reruns and theatrical films to partnering with studios to create new and original series that fit their brand image.[12] In the early 2010s, streaming video portals like Netflix, Hulu, YouTube, and Amazon Video took a similar approach by first building vast libraries of licensed content and then spending on original or exclusive properties. In all cases, companies entered into the television realm by serving as a distributor for someone else's programming before trying to manufacture more control over their schedules or libraries. Twitter and Facebook, then, were attempting a modern take on a familiar pivot. Having facilitated television-related chatter and productivity throughout the Social TV era, Twitter and Facebook were ready to bring the chatter, productivity, and video content all to one place—and under their auspices. And just as cable networks and streaming portals learned about audience taste and habits through licensed programming, Twitter and Facebook could exploit years of user activity data to understand what video content would work best for their new video initiatives.

THE "FAILURE" OF SOCIAL TV

On the one hand, Twitter and Facebook's video plans can be situated as a natural result of Social TV's failures. History already suggests that Twitter and Facebook would pivot from facilitator to licenser to producer, but the industry-wide rush to secure a place in the streaming video ecosystem guaranteed that shift. Traditional cable and satellite providers continue to lose customers by the millions each year.[13] Primetime television viewership, apart from sporting events, drops every year as well.[14] The continued growth in subscribers and spending from Netflix pushed television consumption and the television industry toward on-demand streaming—and away from live, appointment-style viewing.[15] Netflix's growth inspired every other global media conglomerate to develop an in-house streaming video portal.[16] The increase of short- and medium-form video content on Instagram, TikTok, and YouTube produced new generations of celebrity content creators and "influencers," as well as new relationships between Hollywood and Silicon Valley.[17] Even after the remediation efforts of the Social TV era, television has a different position in the even more crowded attention economy. This new position is identifiable in Hollywood and Silicon Valley's modern obsession with *video* over television, and with *subscribers* over viewers.

On the other hand, Twitter and Facebook's remediation—and attempted erasure—of television only underlines that the medium is continuously in the process of remediation. The cycles of remediation-legitimation do not operate linearly or in a vacuum. Technologies constantly and simultaneously remediate one another. The discourses and negotiations surrounding those technologies also converge, conflict, and circle back. Whereas many of the

Social TV initiatives of the 2010s vowed to transform television by making it *more social*, these developments instead strived to make social platforms a lot more like television—no matter what it was called. Twitter and Facebook, alleged new media disruptors, again looked backward to move forward. They remediated an old medium with familiar protocols to expand their cultural footprint and deepen their partnerships with the media industries. And they appropriated discourses and strategies from long before likes, retweets, or second screens. The resilience of remediation-legitimation cycles indicates that the media and tech industries will never stop borrowing from the past or positioning each new product rollout as momentous, disruptive, or liberating for consumers, even in the face of failure.

The Social TV era's utopian promises about new technology's impact on television elided the fact that many of the initiatives intended to construct an aura of immediacy, ephemerality, and community around television in the face of extensive on-demand availability. The promotion of hashtag chyrons, live-tweeting events, and synchronized second-screen apps exemplified an industry-wide interest in legitimating new habits that kept viewers engaged live as often as possible. Gamified check-in platforms even tried to compensate viewers for tuning in with digital stickers and consumer goods. In moments where there was no genuine liveness to be found, like on Amazon Prime's video portal, the studio utilized temporal restrictions and crowd-sourcing to encourage the maximum amount of participation in the Pilot Season process.

While all of these Social TV schemes hailed audiences as active fans, they also positioned that fandom within very particular frameworks. The most dedicated fans participated in live-tweeting sessions with the cast of *Scandal* or hopped over to HBO Connect to chat with stars from *Veep*. They made moral judgments and compared their answers with random celebrities on *The Walking Dead* Story Sync. They competed with fellow Fans and Super Fans to acquire the most digital stickers and the exclusive Guru title on GetGlue. They spread the word about their favorite projects from the latest Pilot Season on social platforms and with fan groups. Altogether, these television viewers were encouraged to affirm their participation in a collective fan experience by downloading more apps, consuming more ancillary content, and evangelizing even more for their favorite programs.

These characteristics of most appropriate or most rewarded fan activity exhibited the media industries' long-standing efforts to define fandom both widely and narrowly as possible. As Social TV products labeled the most basic activities *as* fandom, they also prevented more robust or critical activities. In cases like Story Sync, check-ins, or Pilot Season, companies pushed fans into digital enclosures defined by temporal restrictions and hyper-ephemeral activity. In cases like #TGIT and HBO's social rebrand, fans had more freedom to act and react but saw their multi-screen energy routed through fairly conservative promotional measures. Moreover, this construction of fandom flattened differences among fans, from technological affordance to demographic identity. Social TV initiatives assumed a high degree of connectivity, fluidity, and collectivity, despite the fundamentally personalized and unique circumstances of even those fans with access to the "right" devices or screens. These initiatives also assumed a familiar whiteness and maleness among fans unless, as in the case of #TGIT, the activity and vernacular of Black people could be used to boost the progressive and disruptive bona fides of a veteran broadcaster. This all tells us that the increased visibility of particular fan activities or fan identities in the social media ecosystem does not guarantee greater influence upon corporate decision-making. Hollywood and Silicon Valley will continue to insist that things are different, that everyone is a fan, and we all have access and control. But the framework of what fans, and what fan activities, truly matter will remain generally unchanged.

The Social TV phenomenon was all about controlling attention and making television *an event*, no matter the context. Some of the initiatives were successful in this pursuit of attention, if only for a short time. *Scandal* probably would not have become a cultural and industrial phenomenon without Kerry Washington's live-tweeting of the pilot episode. The fansourcing and faux-participatory discourses of Pilot Season gave Amazon Studios a marker of differentiation in the crowded streaming video marketplace. Story Sync and the various check-in apps had, at least temporarily, user bases in the millions. And live-tweeting has remained a relevant audience practice, particularly for live events like sports, award shows, and political debates. Liveness and immediacy still matter, but Social TV strategies are no longer viewed as innovative or critical maneuvers for the media industries to harness those feelings.

Of course, Social TV did not *save* television or *make it better*. The eventual failure of second-screen apps or check-ins or Pilot Season was, in many ways, a product of unreasonable hype. Networks, studios, and start-ups positioned Social TV as the future to legitimize their investments and products, as well as convince viewers to watch, download, and participate. The press also played a central role in inflating the potential of each new Social TV product. While most potential viewers do not read trade press reports, the coverage of Nielsen Twitter TV Ratings or GetGlue's "most popular" charts helped to perpetuate the idea that new Social TV initiatives were worth following—and, potentially, worth investing in. The mainstream press coverage of #TGIT and Pilot Season bestowed even more legitimacy on those campaigns. This sweeping agenda setting worked to frame the television industry as innovative and forward-thinking, even in the face of failure.

Indeed, the press' short memory regarding each initiative covered in this book demonstrates how vital those institutions are in establishing and re-establishing remediation-legitimation cycles. Notably, trade publications like *Variety, Hollywood Reporter,* and *Deadline* now rarely publish stories about Nielsen's "Social Ratings." They do, however, regularly report on Nielsen's latest ratings data that attempt to track streaming video consumption on services like Netflix.[18] Though there is a familiar debate about the validity of these numbers, or Nielsen's ability to track streaming behavior effectively, executives and the press nonetheless tout them as significant. When there are new platforms and data points to legitimate, the formerly innovative or relevant figures are left by the wayside.

Although Social TV failed to remain a dominant industry concern, the cases spotlighted herein indicated that remediation can generate compelling relationships between old and new media. As the #TGIT campaign showed, the feelings of connectivity and community produced by Twitter can be intensified when they are tethered to a live episode of television and embedded within the primetime schedule. There, the liveness and flow of television help build anticipation and affective responses that are perfect for social productivity and ephemeral engagement of Twitter. Check-in companies likewise tried to create hyper-mediated, gamified experiences that made viewers hyper-conscious of their viewing habits and the technical capability of their mobile devices. But the check-in processes and rewards were still connected to the live, preprogrammed television schedule.

Therefore, it is expected that Twitter and Facebook would distinctively appropriate the features of television and discourses of remediation. To some degree, the efforts to bring even more content to one specific location suggest that social platforms are trying to be *less ephemeral*. Among all the noise and clutter of the digital media ecosystem, Twitter and Facebook want their platforms to generate a type of modern appointment viewing. Video initiatives aim to give users a new reason to visit the platform and, potentially, keep them engaged with the platform for more extended periods. Social platforms are relying on the same kind of strategies that television networks have deployed for decades.

Television networks and media conglomerates did a lot of remediating and appropriating from social platforms in the Social TV era. By trying to become *more social,* television embraced the characteristics of social platforms: connecting, linking, sharing, liking, and engaging. As a result, television grew more ephemeral and more dislocated from a specific distribution channel. But the changes to television programming and network identities do not indicate the failure of Social TV. Instead, they represent an evolution of programming and networks into content, platforms, and portals that still benefit the television industry in the post-Social TV realm.

THE PLATFORMING OF TELEVISION

With television networks and social platforms—and the industries that control them—working to become more like the other, both have been integrated deeper into the fabric of daily life. Writing of YouTube's initial remediation of television, William Uricchio predicted that an "ontological ambivalence" would take hold within the digital ecosystem. Uricchio argues:

> At a moment when the full implications of the digital turn have yet to transform our ways of thinking about moving-image content and our categories of analysis, when the relations between producers and consumers characteristic of the industrial era are slowly being eroded, and when convergent media industries are themselves spreading content across as many platforms as possible, YouTube offers a site of aggregation that exacerbates—and capitalizes upon—that uncertainty.[19]

Uricchio's analysis presumed that YouTube would remix the protocols and discourses of television within a more participatory and user-generated context. Richard Grusin likewise argued that YouTube would function "as a remediation of television in the 'world of network public.'"[20] These predictions have come true. YouTube's enormous growth is just as noteworthy as Netflix's expansion in pushing Hollywood toward video content strategies and streaming video distribution. However, in the process, YouTube has become a far more professionalized space colonized by the traditional media industries. The company has cycled through various premium and subscription tiers of YouTube-exclusive content.[21] But it found its highest degree of success through YouTube TV, an over-the-top streaming video version of conventional cable and satellite service.[22] YouTube, like Twitter and Facebook, has become more like television.

This ontological ambivalence reshaped television throughout the Social TV era. Armed with an array of distribution channels, including in-house streaming portals, and even more social platform accounts, television networks now exist everywhere. As evidenced by the changes in HBO's Twitter persona, networks are more than ephemeral. They are omnipresent. They do not merely distribute programs for live and on-demand consumption, nor do they only exist as brands with images carried out in commercials, logos, and awards legitimation. Instead, they are always accessible from any screen, at any time, and as willing to post a GIF to score 100 retweets as they are to promote the previous night's Nielsen rating. While this perpetual accessibility and immediacy has been inherent to television since its inception, the shift to the digital and social spheres has compounded it immensely. In this regard, the efforts to reaffirm liveness, flow, branding, and repurposing have transformed the ABCs, AMCs, and HBOs of the world into something beyond television networks: platforms.

"Platform," like "network" or "channel," is a structural metaphor that, as Tarleton Gillespie argues, "depends on terms and ideas that are specific enough to mean something, and vague enough to work across multiple venues for multiple audiences."[23] These terms are grounded in technological reality but also used as discursive constructs that shape how people imagine their relationship with technology. In this regard, one could make the case that the concepts of "network" and "channel" have evolved along with broader shifts in technology. Perhaps we could say that network now refers to both an organizer and distributor of television content and the *network* of connected viewers. Likewise, perhaps "channel" helps us describe how viewers are now *channeled* from one content stream to another across the web.

Nevertheless, the platform concept helps designate how television networks explicitly promise to offer viewers the chance to communicate, interact, and participate with television. Networks aim to translate the endless rhetoric about the democratizing potential of the internet and social platforms to television. Viewers are always encouraged to watch, talk back, deliver feedback, and participate in a never-ending and no-longer-time-sensitive discussion about their most beloved or most hated programs. But like Twitter and Facebook, television networks obscure the degree to which they consider this conversation, noting it only when it points to their popularity in newer spaces. Platform, therefore, stresses the multiplicity of constituencies these companies seek to attract. Gillespie writes that platforms must work to "strike a different balance between safe and controversial, between socially and financially valuable, between niche and wide appeal."[24] Television networks have always aimed to reach a diverse range of viewers. Still, the vast expansion of access points has inarguably made that process more challenging and more visible, and thus more urgent. Fostering genuine fans, or at least consistent subscribers, among consumers with so many options requires a never-ending circulation of content that goes beyond a single live episode or a night's worth of programming.

Platform also helps describe how these corporations engage in political culture and commodity activism. While television networks have never been apolitical, broader shifts in American capitalist culture dictate that they now more overtly participate in ongoing conversations or debates that are identified as political. These are moments where changing a Facebook profile to support Pride month or announcing a donation to Black Lives Matter on Instagram is not only newsworthy but cataloged by followers. Participation in this realm works to establish the social personas of television platforms as well as to tell followers that personified corporations are listening to and supporting them. Given the immense politicization of modern consumption, this performance of platform authenticity can help companies score newfound loyalty and subscriptions from different sectors of the potential audience.

Television programs still matter as television networks are platformed, but they are part of a much larger ecosystem of content. While they drive attention and subscriptions on their own, they also serve as generators of social, short-form, and ancillary content: video snippets, memes, trending hashtags, Instagram stories, podcasts, and so on. Again, using Ethan Tussey's term, this "snackable" content helps raise the profile of specific programs and, occasionally, drives viewers to watch in the established live setting.[25] But the shift to content production and circulation just as often manifests as political advocacy or ironic riffing on the viral memes of the day. People post and share GIFs that spotlight famous moments from television series they have never seen. As such, the embrace of content helps programs occupy, or as Tussey writes, "colonize," the timelines, platforms, and attention far beyond normal live viewing.[26] Individual episodes or individual tweets are hyper-ephemeral. The continual and cumulative effect of the distribution of television content is *everyday omnipresence*.

Meanwhile, social platforms play a critical role in the platforming of television. Twitter and Facebook provide the technical infrastructure to facilitate interactions between television corporations and individuals. They promote the circulation of television content across their platforms. As Twitter and Facebook remediate television and integrate television-like strategies into their platforms, they also confirm the platforming of television. Social platforms view television as a way to garner attention from viewers, potential sponsors, and television producers hoping to extend their footprints in the social realms.

The result of television networks operating like platforms and platforms operating like television networks is a perpetual and immediate barrage of content. Circling back to his work on remediation, Grusin presents the concept of "premediation," which is "fostered and encouraged by mobilizing or intensifying pleasurable affects in the production of multiple, overlapping feedback loops among people (individually and collectively) and their media."[27] For Grusin, the intensification and multiplication of media platforms help produce conditions where all future events already feel premediated. Despite the increasing influence of social platforms, television platforms still play a central role in this process. Television platforms consistently circulate content that previews content-to-come, creating an endless flow of material to consume and share. News breaks on network Twitter accounts and is later discussed on television. Promotional clips follow a just-ended episode and then get circulated on Facebook, spoiling future events. Streaming video portals like Netflix and Hulu analyze user tastes and suggest new viewing choices based on necessary algorithmic input. Content is always-already premediated and distributed to those who want it.

More pointedly, the platforming of television represents a greater encroachment of these corporate entities into everyday life. This is a natural acceleration of the movement in the later stages of the twentieth century where global corporations began to plaster previously unoccupied public space with logos and brand iconography.[28] The expansion of "ambient advertising" on gas station pumps, park benches, trash cans, and the like helped normalize consumers' affective relationships with brands.[29] As Anna McCarthy famously detailed, the rise of "ambient television" in settings like airports and waiting rooms had an equally occupying effect on public space.[30] With physical space effectively colonized, corporations have turned their energy to the frontier of platforms, interfaces, and screens, where attention is an unlimited resource to be mined and exploited for capitalist ends.

The Social TV era demonstrated how the line between Michel de Certeau's strategies and tactics continues to blur. For de Certeau, institutions use strategies to structure and control everyday life, while individuals are left with tactics to negotiate those structures and potentially resist that control.[31] Social platforms and digital production tools have enabled individuals to resist the systematizing of the everyday. Still, the corporations behind those platforms and tools have grown accustomed to selling this reality back to individuals as part of participatory and democratizing discourses. Indeed, as Lev Manovich argues, platforms give individuals the tools to develop tactics within branded infrastructures.[32] Forms of participation are funneled through processes of data collection and then employed to facilitate additional forms of engagement. This is key to the television platform strategy, where viewers or users are courted as fans, and their tactics are solicited as meaningful to the production, distribution, and promotion of television. Individuals still manage to build tactical actions in this ecosystem—whether explicitly in the filtering out of content they dislike or implicitly in ignoring strategies like the various failures of Social TV. But they still act within heavily structured platform environments that claim, more than ever, to be unstructured.

I return then to Henri Lefebvre's idea that television has long tried to construct "moments" of everyday life where all events are "presented" as part of a perpetual present and therefore worth the spectacle and the attention.[33] In trying to combat the increasing number of entertainment choices available to multitasking viewers, television networks have turned to platforming as a way to eventize *themselves*, not just their programming. They sell an omnipresent authenticity driven by the technological logic of platforms and the relationship logic of lifestyle branding. The process of making television more social has brought networks, operating as platforms, closer to people. As their ephemeral content circulates, the everyday experience becomes even more of a corporatized, multi-platform spectacle.

CONCLUSION: EPHEMERAL HISTORY

My experience tracking the Social TV phenomenon in many ways mirrored its hype cycle within Hollywood and Silicon Valley. When I first became interested in Social TV in 2012, it was pegged as the future. Nearly every media conglomerate was funding second-screen apps. Check-ins were attracting television fans with rewards. The trade press was reporting on the swell of Twitter chatter surrounding live television events. By 2015, it was clear that television had grown more intertwined with social media. But the excitement surrounding Social TV had subsided as the streaming video competition became the central industry concern. By 2017, Social TV mattered so little that it did not warrant a mention when Twitter and Facebook announced the new video initiatives discussed at the beginning of this chapter. Social TV, a loosely coordinated effort to eventize the ephemerality of television, was just a momentary blip, a micro-moment, in the story of the convergent media industries.

As long as culture remains focused on new technology and techno-utopian discourses, the cycles of hype, remediation, legitimation, and erasure will continue. Micro-moments will come and go faster than ever. The successes and failures within those micro-moments are essential to our understanding of the media industries and failure. Treating recent ephemera as history can be challenging, particularly as the media industries seek to make us forget about the past as they move on to the next big thing of the future. As the Social TV example underlines, individual failures are almost always instructive for future media industry strategy. Social TV struggled to protect live television, but it certainly helped push television further into our lives.

NOTES

INTRODUCTION

1. Jessica Guynn, "Ellen DeGeneres Selfie on Twitter is Oscar Gold for Samsung," *Los Angeles Times*, March 3, 2014, http://articles.latimes.com/2014/mar/03/business/la-fi-tn-ellen-degeneres-selfie-twitter-oscars-samsung-20140303.

2. "Oscars Selfie," *Time 100 Photos*, n.d., http://100photos.time.com/photos/bradley-cooper-oscars-selfie.

3. Fred Graver, "The True Story of the 'Ellen Selfie,'" *Medium*, February 23, 2017, https://medium.com/@fredgraver/the-true-story-of-the-ellen-selfie-eb8035c9b34d.

4. Suzanne Vranica, "Behind the Preplanned Oscar Selfie: Samsung's Ad Strategy," *Wall Street Journal*, March 3, 2014, https://www.wsj.com/articles/behind-oscar-selfie-samsungs-ad-strategy-1393886249.

5. Lisa de Moraes, "Is Ellen DeGeneres' Oscar Selfie A Game-Changer for Trophy Show Product Placement?" *Deadline*, March 3, 2014, http://deadline.com/2014/03/ellen-degeneres-oscar-selfie-twitter-product-placement-692969/.

6. I will refer to audiences, fans, users, and viewers. There are particular slippages between the terms in a multi-screen context, with people serving as both viewers of programming and users of social platforms. I also acknowledge that these industries construct audiences *as* fans.

7. Tony Maglio, "Oscars Propel ABC to Most-Watched Week in 4 Years," *The Wrap*, March 4, 2014, https://www.thewrap.com/oscars-propel-abc-watched-week-4-years/.

8. I use "networks" to refer to both conventional broadcast networks (i.e., ABC or CBS) and cable outfits (i.e., AMC or ESPN). While cable channel is the more historically appropriate term, network reflects evolutions in how US television is organized and functions efficiently in the text.

9. Amy Jo Coffey, "Promotional Practices of Cable Networks: A Comparative Analysis of New and Traditional Spaces," *International Journal on Media Management* 13, no. 3 (2011): 161–76; Ruth Deller, "Twittering On: Audience Research and Participation Using Twitter," *Participations: Journal of Audience & Reception Studies* 8, no. 1 (2011): 216–45; Clark F. Greer and Douglas A. Ferguson, "Using Twitter for Promotion and Branding: A Content Analysis of Local Television Twitter Sites," *Journal of Broadcasting & Electronic*

Media, 55, no. 2 (2011): 198–214; D. Yvette Wohn and Eun-Kyung Na, "Tweeting about TV: Sharing Television Experiences Via Social Media Message Streams," *First Monday* 16, no. 3 (2011), https://firstmonday.org/article/view/3368/2779; Megan M. Wood and Linda Baughman, "*Glee* Fandom and Twitter: Something New, or More of the Same Old Thing?" *Communication Studies* 63, no. 3 (2012): 328–44; Stephen Harrington, Tim Highfield, and Axel Bruns, "More than a Backchannel: Twitter and Television," *Participations* 10, no. 1 (2013): 405–9.

10. Hye Jin Lee and Mark Andrejevic, "Second-Screen Theory: From the Democratic Surround to the Digital Enclosure," in *Connected Viewing: Selling, Streaming, & Sharing Media in the Digital Era*, eds. Jennifer Holt and Kevin Sanson (New York: Routledge, 2014).

11. Raymond Williams, *Television: Technology and Cultural Form* (London: Routledge, 1990), 79–85.

12. Holt and Sanson, "Introduction: Mapping Connections," in *Connected Viewing: Selling, Streaming, & Sharing Media in the Digital Era*, eds. Jennifer Holt and Kevin Sanson (New York: Routledge, 2014), 1–15; Matthew Pittman, and Alec C. Tefertiller, "Without or Without You: Connected Viewing and Co-Viewing Twitter Activity for Traditional Appointment and Asynchronous Broadcast Television Models., *First Monday* 20, no. 7 (2015), http://firstmonday.org/ojs/index.php/fm/article/view/5935.

13. "Disruption" originally comes from Clayton Christensen, *The Innovator's Dilemma: When New Technologies Cause Great Firms to Fail* (New York: Harvard Business Review Press, 1997), 3–10. See also, "'Silicon Valley' Asks: Is Your Start-up Really Making the World A Better Place?" National Public Radio, April 17, 2014, https://www.npr.org/2014/04/17/304150243/silicon-valley-asks-is-your-start-up-really-making-the-world-better.

14. Andy Rachleff, "What 'Disrupt' Really Means," *TechCrunch*, February 16, 2013, https://techcrunch.com/2013/02/16/the-truth-about-disruption/. For more detailed popular accounts of "disruption," see Walter Isaacson, "How Uber and Airbnb Became Poster Children for the Disruption Economy," *New York Times*, June 19, 2017, https://www.nytimes.com/2017/06/19/books/review/wild-ride-adam-lashinsky-uber-airbnb.html; and Lee Fang, "Venture Capitalists Are Poised to 'Disrupt' Everything about the Education Market," *The Atlantic*, September 25, 2014, https://www.thenation.com/article/venture-capitalists-are-poised-disrupt-everything-about-education-market/.

15. James Carey and John J. Quirk, "The Mythos of the Electronic Revolution," in *Communication as Culture: Essays on Media and Society*, ed. James Carey (New York: Routledge, 1989), 113–41; James Hughes, *Citizen Cyborg: Why Democratic Societies Must Respond to the Redesigned Human of the Future* (Cambridge: Basic Books, 2004).

16. Jack Smith IV, "'Silicon Valley' Fact Check: HBO Nails TechCrunch Disrupt All the Way Down to the Nametags," *Beta Beat*, June 2, 2014, http://betabeat.com/2014/06/silicon-valley-fact-check-hbo-nails-techcrunch-disrupt-all-the-way-down-to-the-nametags/.

17. Jay Bolter and Richard Grusin, *Remediation: Understanding New Media* (Cambridge: MIT Press, 1999), 55.

18. Bolter and Grusin, *Remediation*, 60.

19. Lisa Gitelman, *Always Already New: Media History, and the Data of Culture* (Cambridge: MIT Press, 2008), 6; William Boddy, *New Media and Popular Imagination*, (Oxford: Oxford University Press, 2004), 1.

20. danah boyd, "Social Media is Here to Stay ... Now What?" Microsoft Research Tech Fest, Redmond, Washington, February 26, 2009, http://www.danah.org/papers/talks/MSRTechFest2009.html.

21. Mark Andrejevic, "Exploitation in the Data Mine," in *Internet and Surveillance: The Challenges of Web 2.0 and Social Media*, eds. Christian Fuchs et al. (New York: Routledge, 2012).

22. Ralph Lee Smith, *The Wire Nation: Cable TV, The Electronic Communications Highway* (New York: Harper & Row, 1972); Jennifer S. Light, "Before the Internet, There Was Cable," *IEEE Annals of the History of Computing* 25, no. 2 (2003): 96.

23. Max Dawson, *TV Repair: New Media 'Solutions' to Old Media Problems* (PhD dissertation, Northwestern University, 2008), 49–110; Vance Packard, *The Hidden Persuaders* (New York: IG Publishing, 2007 [1957]).

24. Dawson, *TV Repair*, 179–80.

25. Nicolas Negroponte, *Being Digital* (London: Hodder and Stoughton, 1995), 19.

26. George Gilder, Life After Television: The Coming Transformation of Media and American Life (New York: Norton, 1994), 40–42.

27. Phillip Swan, *TVDotCom: The Future of Interactive Television* (New York: TV Books, 2000), 16.

28. Sally Bedell Smith, "Who's Watching TV? It's Getting Hard to Tell," *New York Times*, January 6, 1985, H23; Victor Livingston, "Statistical Skirmish: Nielsen Cable Stats Vex Cable Net Execs," *Television/Radio Age*, March 17, 1986, 130.

29. Beville, *Audience Ratings*.

30. Ien Ang, *Desperately Seeking the Audience* (London: Routledge, 1991), 61.

31. Ang, *Desperately Seeking the Audience*, 60.

32. Jennifer Hessler, "Peoplemeter Technologies and the Biometric Turn in Audience Measurement," *Television & New Media* 22, no. 4 (2021): 400–419.

33. "Television in the People Meter," *Broadcasting*, September 7, 1987; "Network Researchers Rate People Meters," *Broadcasting*, April 6, 1987; "NAB, Networks Call for Study of People Meters," *Broadcasting*, March 21, 1988.

34. "TiVo (2000)," YouTube, January 29, 2013, https://www.youtube.com/watch?v=f2X1p0sX9CM.

35. Amy Harmon, "Skip-the-Ads TV Has Madison Ave. Upset," *New York Times*, May 23, 2002, http://www.nytimes.com/2002/05/23/business/skip-the-ads-tv-has-madison-ave-upset.html.

36. Ken Fisher, "ReplayTV Lawsuit, Examined," *Ars Technica*, November 13, 2001, https://arstechnica.com/uncategorized/2001/11/2551-2/.

37. "Future Shock," *Mediaweek*, May 10, 2004; "Measuring Up: VOD Still a Gray Area for Nielsen," *Cablefax*, April 27, 2004; David Bauder, "For Nielsen Ratings, Problems with Networks Are a Familiar Story," Associated Press, November 14, 2003.

38. Brent Sporich, "TiVo, Nielsen Will Scan DVR Time-Shift Patterns," *Hollywood Reporter*, July 26, 2000; Meredith Amdur and Pamela McClintock, "Commercial Free?"

Variety, February 5, 2004, 8; Stuart Elliott, "How to Value Ratings with DVR Delay," *New York Times*, February 13, 2006, C1, 15.

39. John T. Caldwell, "Second-Shift Media Aesthetics: Programming, Interactivity, and User Flows," in *New Media: Theories and Practices of Digitextuality*, eds. Anna Everett and John T. Caldwell (New York: Routledge, 2003), 127–44.

40. Max Dawson, "Little Players, Big Shows: Format, Narrative, and Style on Television's New Smaller Screens," *Convergence* 13, no. 3 (2007): 231–50; Jennifer Gillan, *Television and New Media: Must-Click TV* (New York: Routledge, 2011).

41. Portals comes from Amanda Lotz's theorization of the streaming ecosystem. Amanda Lotz, *Portals: A Treatise on Internet-Distributed Television* (Ann Arbor: University of Michigan Publishing, 2017). For more on executive predictions, see Daisy Whitney, "New Order, New Champs: Networks Learning Which Shows Flourish with New Media Exposure—and Which Don't," *Television Week*, March 27, 2006, 3; Paul J. Gough, "Online Streams Help ABC Affils," *Hollywood Reporter*, April 26, 2006.

42. Nielsen, "10 Years of Primetime: The Rise of Reality and Sports Programming," *Nielsen Insights*, September 21, 2011, https://www.nielsen.com/us/en/insights/article/2011/10-years-of-primetime-the-rise-of-reality-and-sports-programming/; Bill Carter, "Prime-Time Ratings Bring Speculation of a Shift in Habits," *New York Times*, April 22, 2012, https://www.nytimes.com/2012/04/23/business/media/tv-viewers-are-missing-in-action.html.

43. Ryan Lawler, "Cord Cutters Are Young, Educated, and Employed," *Gigaom*, October 29, 2010, https://gigaom.com/2010/10/29/cord-cutters-are-young-educated-and-employed/; Shalini Ramachandran, "Evidence Grows on TV Cord-Cutting," *Wall Street Journal*, August 7, 2012, https://www.wsj.com/articles/SB10000872396390443792604577574901875760374.

44. Annie Barrett, "Twitter on Fox: Why Does My TV Look Like A Computer?" *Entertainment Weekly*, September 4, 2009, http://ew.com/article/2009/09/04/twitter-on-fox-why-does-my-tv-look-like-a-computer/.

45. Gillan, *Must-Click TV*, 234–36.

46. Mandi Bierly, "'Community' Exclusive: Twittersode to Precede Season Premiere," *Entertainment Weekly*, September 21, 2010, http://ew.com/article/2010/09/21/community-twittersode/.

47. Michael Schneider, "New to Your TV Screen: Twitter and Hashtags," *TV Guide*, April 21, 2011, http://www.tvguide.com/news/new-tv-screen-1032111/.

48. David Goetzl, "CBS Launches Tweet Week," *MediaPost*, March 31, 2011, https://www.mediapost.com/publications/article/147835/cbs-launches-tweet-week.html.

49. Biz Stone, "Twitter Goes Hollywood?" *Twitter Blog*, May 26, 2009, https://blog.twitter.com/official/en_us/a/2009/twitter-goes-hollywood.html.

50. "#superbowl," *Twitter Blog*, February 9, 2011, https://blog.twitter.com/official/en_us/a/2011/superbowl.html.

51. "Sunday Night TV on Twitter," *Twitter Blog*, April 17, 2012, https://blog.twitter.com/official/en_us/a/2012/sunday-night-tv-on-twitter.html.

52. Ingrid Lunden, "Twitter Launches TV Ad Targeting, Twitter Amplify for Real-Time Video Stream," *TechCrunch*, May 23, 2013, https://techcrunch.com/2013/05/23/twitter

-launches-twitter-amplify-for-real-time-videos-in-stream-partnering-with-bbc-fox-fuse-and-weather-channel/.

53. "Twitter, TV, and You," *Twitter Blog*, October 26, 2011, https://blog.twitter.com/official/en_us/a/2011/twitter-tv-and-you.html.

54. Radha Subramanyam, "The Relationship Between Social Media Buzz and TV Ratings," *Nielsen*, October 6, 2011, http://www.nielsen.com/us/en/insights/news/2011/the-relationship-between-social-media-buzz-and-tv-ratings.html.

55. "Social Media and TV—Who's Talking When and What About," *Nielsen*, October 11, 2011, http://www.nielsen.com/us/en/insights/news/2011/social-media-and-tv-whos-talking-when-and-what-about.html.

56. Bob Fernandez, "Comcast and NBCUniversal Take a Stake in British Social TV Start-Up," *Philadelphia Inquirer*, September 27, 2012; Samantha Murphy, "More Americans Are Using Mobile Phones While Watching TV," *Mashable*, July 17, 2012, http://mashable.com/2012/07/17/mobile-phones-tv/; Cory Bergman, "Study: 63 Percent of Tablet Owners Multitask with TV," *Lost Remote*, November 7, 2012, http://lostremote.com/study-63-percent-of-tablet-owners-multitask-with-tv_b34956.

57. Rick Kissell, "Social TV Chatter Grows 800% on Twitter over 2012," *Variety*, December 26, 2012, http://variety.com/2012/tv/news/social-tv-chatter-grows-800-on-twitter-over-2012-1118063975/.

58. "Nielsen and Twitter Establish Social TV Rating," *Nielsen*, December 17, 2012, http://www.nielsen.com/us/en/insights/press-room/2012/nielsen-and-twitter-establish-social-tv-rating.html.

59. Sarah Banet-Weiser, *Authentic (TM): The Politics of Ambivalence in a Brand Culture* (New York: NYU Press, 2012).

60. Ethan Tussey, *The Procrastination Economy: The Big Business of Downtime* (New York: NYU Press, 2018), 30.

61. Lynn Spigel, *Make Room for TV: The Family Ideal in Postwar America* (Chicago: University of Chicago Press, 1992).

62. For examples of this television brand analysis, see Amanda D. Lotz, *Redesigning Women: Television After the Network Era* (Urbana: University of Illinois Press, 2006); Barbara Selznick, "Branding the Future: Syfy in the Post-Network Era," *Science Fiction Film and Television* 2, no. 2 (2009): 177–204; Anthony N. Smith, "Putting the Premium into Basic: Slow-Burn Narratives and the Loss-Leader Function of AMC's Original Drama Series," *Television & New Media* 14, no. 2 (2012): 150–66; and Anthony N. Smith, "Pursuing 'Generation Snowflake': *Mr. Robot* and the USA Network's Mission for Millennials," *Television & New Media* 20, no. 5 (2019): 443–59.

63. Catherine Johnson, *Branding Television* (New York: Routledge, 2012), 2–4; Banet-Weiser, *Authentic*, 9. See also Adam Arvidsson, *Brands: Meaning and Value in Media Culture* (New York: Routledge, 2006) and Celia Lury, *Brands: The Logos of the Global Economy* (New York: Routledge, 2004).

64. Paul Grainge, "Introduction: Ephemeral Media," in *Ephemeral Media: Transitory Screen Culture from Television to YouTube*, ed. Paul Grainge (London: British Film Institute, 2011), 2.

65. Grainge, "Introduction: Ephemereal Media," 3.

66. Dawson, "Little Players, Big Shows."

67. Steven M. Schneider and Kirsten Foot, "The Web as an Object of Study," *New Media & Society* 6, no. 1 (2004): 115.

68. Amelie, Hastie, "Detritus and the Moving Image: Ephemera, Materiality, History," *Journal of Visual Culture* 6, no. 2 (2007): 171–74.

69. Elizabeth Evans, "'Carnaby Street, 10am': *KateModern* and the Ephemeral Dynamics of Online Drama," in *Ephemeral Media: Transitory Screen Culture from Television to YouTube*, ed. Paul Grainge (London: British Film Institute, 2011), 159.

70. Helen Wheatley, "Introduction: Re-Viewing Television Histories," in *Re-Viewing Television Histories: Critical Issues in Television Historiography*, ed. Helen Wheatley (New York: I. B. Tauris, 2007), 3.

71. Rick Altman, *Silent Film Sound* (New York: Columbia University Press, 2005), 16.

72. Altman, *Silent Film Sound*, 17.

73. Gill Branston, "Histories of British Television," in *The Television Studies Book*, eds. Christine Geraghty and David Lusted (London: Arnold, 1998), 53; Richard Johnson, "Historical Returns: Transdisciplinarity, Cultural Studies, and History," *European Journal of Cultural Studies* 4, no. 3 (2001): 279; Wheatley, "Re-Viewing Television Histories," 3.

74. Jean Burgess and Joshua Green, *YouTube: Online Video and Participatory Culture* (Cambridge: Polity Press, 2009); Steve F. Anderson, *Technologies of History: Visual Media and the Eccentricity of the Past* (Hanover: Dartmouth College Press, 2011).

75. Jonathan Gray, *Shows Sold Separately: Promos, Spoilers, and Other Media Paratexts* (New York: NYU Press, 2010), 4–8.

76. Gray, *Shows Sold Separately*, 4–10. Gray draws from Gerard Genette, *Paratexts: Thresholds of Interpretation*, trans. Jane E. Lewin (Cambridge: Cambridge University Press, 1997).

77. Karin van Es, "Social TV and the Participation Dilemma in NBC's *The Voice*," *Television & New Media* 17, no. 2 (2016): 108–23; Sherryl Wilson, "In the Living Room: Second Screens and TV Audiences," *Television & New Media* 17, no. 2 (2016): 174–91.

78. For examples of scholars calling for a refocus or evolution of audience-oriented work, see Jonathan Gray, "Reviving Audience Studies," *Critical Studies in Media Communication* 34, no. 1 (2017): 79–83; Sonia Livingstone, "Audiences in the Age of Datafication: Critical Questions for Media Research," *Television & New Media* 20, no. 2 (2019): 170–83; Graeme Turner, "Approaching the Cultures of Use: Netflix, Disruption, and the Audience," *Critical Studies in Television* 14, no. 2 (2019): 222–32.

79. John T. Caldwell, *Production Culture: Industrial Reflexivity and Critical Practice in Film and Television* (Durham: Duke University Press, 2008); Alisa Perren and Jennifer Holt, eds., *Media Industries Studies: History, Theory, and Method* (New York: Wiley, 2011).

80. Alex Weprin, Much A-Twitter About Something," *Broadcasting & Cable*, May 11, 2009, 12–13.

81. Stuart Levine and Cynthia Littleton, "TV Creators Ace Face Time," *Variety*, April 5, 2010, 1–2.

82. Sharon M. Ross, *Beyond the Box: Television and the Internet* (Malden: Wiley-Blackwell, 2008), 53–55.

83. Henry Jenkins, *Convergence Culture: Where Old and New Media Collide* (New York: NYU Press, 2006).

84. Simone Murray, "'Celebrating the Story the Way It Is': Cultural Studies, Corporate Media and the Contested Utility of Fandom," *Continuum* 18, no. 1 (2004): 7–25.

85. Paul Booth, *Playing Fans: Negotiating Fandom and Media in the Digital Age* (Iowa City: University of Iowa Press, 2015), 120.

86. Mel Stanfill, *Exploiting Fandom: How the Media Industry Seeks to Manipulate Fans* (Iowa City: University of Iowa Press, 2019), 11.

87. Louisa Ellen Stein, *Millennial Fandom: Television Audiences in the Transmedia Age* (Iowa City: University of Iowa Press, 2015), 4.

88. Stanfill, *Exploiting Fandom*, 32–34.

89. Suzanne Scott, *Fake Geek Girls: Fandom, Gender, and the Convergence Culture Industry* (New York: NYU Press, 2019).

90. Kristen Warner, "ABC's *Scandal* and Black Women's Fandom," in *Cupcakes, Pinterest, and Ladyporn: Feminized Popular Culture in the Early Twenty-First Century*, ed. Elana Levine (Urbana: University of Illinois Press, 2015), 32–50; Rebecca Wanzo, "African American Acafandom and Other Strangers: New Genealogies of Fan Studies," *Transformative Works and Cultures*, no. 20 (2015), https://doi.org/10.3983/twc.2015.0699.

91. Michael Z. Newman and Elana Levine, *Legitimating Television: Media Convergence and Cultural Status* (New York: Routledge, 2011).

92. Deborah L. Jaramillo, "The Family Racket: AOL-Time Warner, HBO, *The Sopranos*, and the Construction of a Quality Brand," *Journal of Communication Inquiry* 26, no. 1 (2002): 59–75; Jane Feuer, "HBO and the Concept of Quality TV," in *Quality TV: Contemporary American Television and Beyond*, eds. Janet McCabe and Kim Akass (New York: I. B. Tauris, 2007), 145–57; Avi Santo, "Para-Television and Discourses of Distinction: The Culture of Production at HBO," in *It's Not TV: Watching HBO in the Post-Television Era*, eds. Marc Leverette, Brian L. Ott, and Cara Louise Buckley (New York: Routledge, 2008), 19–43; Taylor Nygaard, "Girls Just Want to be "Quality": HBO, Lena Dunham, and *Girls*' Conflicting Brand Identity," *Feminist Media Studies* 13, no. 2 (2013): 370–74.

93. Taylor Nygaard and Jorie Lagerwey, "Broadcasting Quality: Re-Centering Feminist Discourse with *The Good Wife*," *Television & New Media* 18, no. 2 (2017): 105–13.

94. Scott, *Fake Geek Girls*, 3.

95. Sarah Florini, *Beyond Hashtags: Racial Politics and Black Digital Networks* (New York: NYU Press, 2019); André Brock Jr., *Distributed Blackness: African American Cybercultures* (New York: NYU Press, 2020).

96. Kristina Busse, "Geek Hierarchies, Boundary Policing, and the Gendering of the Good Fan," *Participations* 10, no. 1 (2013): 77; Will Brooker, "Going Pro: Gendered Responses to the Incorporation of Fan Labor as User-Generated Content," in *Wired TV: Laboring Over an Interactive Future*, ed. Denise Mann (New Brunswick: Rutgers University Press, 2014), 76; Cory Barker, "'Social TV': *Pretty Little Liars*, Casual Fandom, Celebrity Instagramming, and Media Life," *Journal of Popular Culture Studies* 2, nos. 1–2 (2014): 215–42; Chuck Tryon, *On-Demand Culture: Digital Delivery and the Future of Movies* (New Brunswick: Rutgers University Press, 2013), 120; Vincent Miller, "New Media, Networking, and Phatic Culture," *Convergence* 14, no. 4 (2008): 387–400.

97. Tarleton Gillespie, "The Politics of Platforms," *New Media & Society* 12, no. 3 (2010): 351–52.

98. José van Dijck, *Culture of Connectivity: A Critical History of Social Media* (Oxford: Oxford University Press, 2013), 31.

99. Lev Manovich, *The Language of New Media* (Cambridge: MIT Press, 2001), 77–79; Alexander Galloway, *The Interface Effect* (New York: Polity Press, 2012), vii.

100. Laura Roman, "Question for Free Chicken Nuggets Inspires Twitter's Most Retweeted Tweet," National Public Radio, May 9, 2017, https://www.npr.org/sections/the-two-way/2017/05/09/527597422/quest-for-free-chicken-nuggets-inspires-twitters-most-retweeted-tweet.

101. *The Ellen Show*, "Ellen Meets Her Chicken Nugget Twitter Opponent," YouTube, April 18, 2017, https://www.youtube.com/watch?v=wiGZgdFHSSc.

102. *Today*, "Wendy's Nugget Teen Carter Wilkerson Beats Ellen DeGeneres's Retweet Record," YouTube, May 10, 2017, https://www.youtube.com/watch?v=4k1JzcESoGo&t=2s.

CHAPTER 1

1. Bill Keveney, "Express Your #TGIT Delight in Emoji Form," *USA Today*, September 24, 2015, http://www.usatoday.com/story/life/entertainthis/2015/09/24/express-your-#TGIT-delight-emoji-form/72739386/.

2. Lauren Johnson, "Twitter's Branded Emojis Come with a Million-Dollar Commitment," *Adweek*, February 2, 2016, http://www.adweek.com/digital/twitters-branded-emojis-come-million-dollar-commitment-169327/.

3. Keveney, "Express Your #TGIT Delight."

4. SonofTheBronx, "*Empire* Season Premiere Tops Week's Series and Specials on Twitter," *TV Media Insights*, September 29, 2015, http://www.tvmediainsights.com/social-tv-buzz/empire-season-premiere-tops-week-series-and-specials-on-twitter/.

5. Shelli Weinstein, "How *Scandal* Paved the Way for ABC's Twitter-Based '#TGIT' Marketing Strategy," *Variety*, September 22, 2014, http://variety.com/2014/tv/news/scandal-twitter-shonda-rhimes-#TGIT-abc-shondaland-1201311282/.

6. Jennifer Gillan, *Television Brandcasting: The Return of the Content-Promotion Hybrid* (New York: Routledge, 2014); Cory Barker, "'Social TV': *Pretty Little Liars*, Casual Fandom, Celebrity Instagramming, and Media Life," *Journal of Popular Culture Studies* 2, nos. 1–2 (2014): 215–42; Melanie E. S. Kohnen, "Cultural Diversity as Brand Management in Cable Television," *Media Industries Journal* 2, no. 2 (2015), https://doi.org/10.3998/mij.15031809.0002.205.

7. Brian Steinberg, "TV Industry Struggles to Agree on Ratings Innovation," *Variety*, April 11, 2017, http://variety.com/2017/tv/features/nielsen-total-content-ratings-1202027752/.

8. Matthew Pittman, and Alec C. Tefertiller, "Without or Without You: Connected Viewing and Co-Viewing Twitter Activity for Traditional Appointment and Asynchronous Broadcast Television Models," *First Monday* 20, no. 7 (2015), http://firstmonday.org/ojs/index.php/fm/article/view/5935.

9. Ruth Deller, "Twittering On: Audience Research and Participation Using Twitter," *Participations* 8, no. 1 (2011): 223.

10. Daniel Dayan and Elihu Katz, *Media Events: The Live Broadcasting of History* (Cambridge: Harvard University Press, 1994).

11. Jérôme Bourdon, "Live Television Is Still Alive: On Television as An Unfulfilled Promised," *Media, Culture & Society*, no. 22 (2000): 550.

12. José van Dijck, *The Culture of Connectivity: A Critical History of Social Media* (Oxford: Oxford University Press, 2013), 78.

13. Nick Couldry, *Media Rituals: A Critical Approach* (London: Routledge, 2003), 97.

14. Stuart Levine and Cynthia Littleton, "TV Creators Ace Face Time," *Variety*, April 5, 2010, 1–2.

15. Michael Z. Newman and Elana Levine, *Legitimating Television: Media Convergence and Cultural Status* (New York: Routledge, 2011), 38–40.

16. Scott Collins, "*Sons of Anarchy*'s Kurt Sutter Lives Up to His 'Wild One' Reputation," *Los Angeles Times*, September 5, 2010, http://articles.latimes.com/2010/sep/05/entertainment/la-ca-kurt-sutter-20100905; Myles McNutt, "Tweets of Anarchy: Showrunners on Twitter," *Antenna*, September 17, 2010, http://blog.commarts.wisc.edu/2010/09/17/tweets-of-anarchy-showrunners-on-twitter/.

17. Alan Wexelblat, "An Auteur in the Age of the Internet: JMS, *Babylon 5*, and the Net," in *Hop on Pop: The Politics and Pleasures of Popular Culture*, eds. Henry Jenkins, Tara McPherson, and Jane Shattuc (Durham: Duke University Press, 2002), 209–26.

18. Suzanne Scott, "Who's Steering the Mothership? The Role of the Fanboy Auteur in Transmedia Storytelling," in *The Participatory Cultures Handbook*, eds. Aaron Delwiche and Jennifer Jacobs Henderson (New York: Routledge, 2013), 43–53.

19. Michaela D. E. Meyer and Rachel Alicia Griffin, "Riding Shondaland's Rollercoasters: Critical Cultural Television Studies in the 21st Century," in *Adventures in Shondaland: Identity Politics and the Power of Representation*, eds. Michaela D. E. Meyer and Rachel Alicia Griffin (New Brunswick: Rutgers University Press, 2018), 5.

20. Kristen Warner, *The Cultural Politics of Colorblind TV Casting* (New York: Routledge, 2015).

21. Lola Ogunnaike, "'Grey's Anatomy' Creator Finds Success in Surgery," *New York Times*, September 28, 2006, http://www.nytimes.com/2006/09/28/arts/television/28anat.html.

22. Scott, "Who's Steering the Mothership?"; Leora Hadas, "A New Vision: J. J. Abrams, *Star Trek*, and Promotional Authorship," *Cinema Journal* 56, no. 2 (2017): 46–66; Anastasia Salter and Mel Stanfill, *A Portrait of the Auteur as Fanboy: The Construction of Authorship in Transmedia Franchises* (Jackson: University Press of Mississippi, 2020).

23. Taylor Nygaard, "Girls Just Want to be 'Quality': HBO, Lena Dunham, and *Girls*' Conflicting Brand Identity," *Feminist Media Studies* 13, no. 2 (2013): 370–74.

24. Matthew Fogel, "'Grey's Anatomy' Goes Colorblind," *New York Times*, May 8, 2005, http://www.nytimes.com/2005/05/08/arts/television/greys-anatomy-goes-colorblind.html.

25. Fogel, "'Grey's Anaomy' Goes Colorblind."

26. Fogel, "'Grey's Anaomy' Goes Colorblind."

27. Pamela K. Johnson, "The Cutting Edge: Shonda Rhimes Dissects *Grey's Anatomy*," *Written By: The Magazine of the Writers Guild of America, West*, September 2006. See also, Aldore Collier, "Shonda Rhimes: The Force Behind *Grey's Anatomy*," *Ebony* 60, no. 12 (2005): 204–08.

28. Newman and Levine, *Legitimating Television*, 48.

29. Herman Gray, *Watching Race: Television and the Struggle for Blackness* (Minneapolis: University of Minnesota Press, 2004).

30. Gray, *Watching Race*, 166.

31. Kristal Brent Zook, *Color by Fox: The Fox Network and the Revolution in Black Television* (New York: Oxford University Press, 1999), 100–102.

32. Beretta E. Smith-Shomade, *Shaded Lives: African-American and Television* (New Brunswick: Rutgers University Press, 2002), 36–37.

33. Ralina L. Joseph, "Strategically Ambiguous Shonda Rhimes: Respectability Politics of a Black Woman Showrunner," *Souls* 18, nos. 2–4 (2016): 306–07.

34. Warner, "The Racial Logic of *Grey's Anatomy*: Shonda Rhimes and Her 'Post-Civil Rights, Post-Feminist' Series," *Television & New Media* 16, no. 7 (2015): 640–44.

35. Lisa Schmidt, "Monstrous Melodrama: Expanding the Scope of Melodramatic Identification to Interpret Negative Fan Responses to *Supernatural*," *Transformative Works and Cultures*, no. 4 (2010), https://doi.org/10.3983/twc.2010.0152. The term "fanboy auteur" comes from Scott, "Who's Steering the Mothership?"

36. "One-on-One with Shonda Rhimes," *Good Morning America*, ABC, September 18, 2014. Cited in Anna Everett, "Scandalicious: *Scandal*, Social Media, and Shonda Rhimes's Auteurist Juggernaut," *Black Scholar* 45, no. 1 (2015): 34–43.

37. Maryann Erigha, "Shonda Rhimes, *Scandal*, and the Politics of Crossing Over," *Black Scholar* 45, no. 1 (2015): 11.

38. Jane Feuer, "Live Television: The Concept of Ontology as Ideology," in *Regarding Television: Critical Approaches*, ed. E. Ann Kaplan (New York: University Publications of America, 1983), 13.

39. John Ellis, *Visible Fictions: Cinema: Television: Video* (New York: Routledge, 1982), 132.

40. Philip Auslander, *Liveness: Performance in a Mediatized Culture, Second Edition* (London: Routledge, 2008), 16; Nick Couldry, "Liveness, 'Reality,' and the Mediated Habitus from Television to the Mobile Phone," *Communication Review* 7, no. 4 (2004): 356–57.

41. Elana Levine, "Distinguishing Television: The Changing Meanings of Television Liveness," *Media, Culture & Society* 30, no. 3 (2008): 395.

42. Raymond Williams, *Television: Technology and Cultural Form* (London: Routledge, 1990), 79.

43. Paddy Scannell, *Television and the Meaning of 'Live': An Inquiry into the Human Situation* (London: Polity Press, 2013), 177–99.

44. Auslander, *Liveness*, 10–12.

45. Ien Ang, *Desperately Seeking the Audience* (London: Routledge, 1991), 15. Ang cites Chris Rojek, *Capitalism and Leisure Theory* (London: Tavistock, 1985), 154–55.

46. Robert Vianello, "The Power Politics of 'Live' Television," *Journal of Film and Video* 37, no. 3 (1985): 28.

47. Levine, "Distinguishing Television," 396–405.

48. Feuer, "Live Television," 15.
49. Williams, *Television: Technology and Cultural Form*, 79–85.
50. Williams, *Television: Technology and Cultural Form*, 93.
51. John Ellis, "Scheduling: The Last Creative Act in Television?" *Media, Culture & Society* 22, no. 1 (2000): 25–38.
52. Uricchio, "Television's Next Great Generation," 170.
53. Uricchio, "Television's Next Great Generation," 184.
54. Shawna Malcolm, "'Grey's Anatomy,' 'Private Practice' Prepare for Joint Operation," *Los Angeles Times*, February 8, 2009, http://articles.latimes.com/2009/feb/08/entertainment/ca-greys-anatomy8.
55. Frazier Moore, "TV's Double Vision, When 1 Screen Isn't Enough," Associated Press, May 3, 2012.
56. Allison Samuels, "How *Scandal* on ABC Got Off the Ground," *Newsweek*, March 5, 2012, http://www.newsweek.com/how-scandal-abc-got-ground-63697.
57. Weinstein, "How 'Scandal' Paved the Way."
58. "Who Is Quinn Perkins?" n.d., http://www.whoisquinnperkins.com/.
59. Max Dawson, "Television's Aesthetic of Efficiency: Convergence Television and the Digital Short," in *Television as Digital Media*, eds. James Bennett and Niki Strange (Durham: Duke University Press, 2011), 204–29.
60. Amanda Kondolojy, "Thursday Final Ratings," *TV by the Numbers*, September 25, 2012, http://tvbythenumbers.zap2it.com/2012/09/28/thursday-final-ratings-big-bang-theory-greys-anatomy-adjusted-up-parks-rec-up-all-night-snl-weekend-update-the-office-glee-scandal-rock-center-adjust/150556.
61. "ABC Launches the #WhoShotFitz Campaign," *TV by the Numbers*, December 4, 2012, http://tvbythenumbers.zap2it.com/network-press-releases/abc-launches-the-official-whoshotfitz-campaign-on-scandal/.
62. van Dijck, *Culture of Connectivity*, 78.
63. Adam Kepler, "Mining 'Scandal' for Ratings," *New York Times*, December 14, 2012, https://artsbeat.blogs.nytimes.com/2012/12/14/mining-scandal-for-ratings/.
64. Tanzina Vega, "A Show Makes Friends and History," *New York Times*, January 16, 2013, http://www.nytimes.com./2013/01/17/arts/television/scandal-on-abc-is-breaking-barriers.html.
65. Lesley Goldberg, "How ABC's 'Scandal' Gets 2,220 Tweets Per Minute," *Hollywood Reporter*, February 7, 2013, https://www.hollywoodreporter.com/news/kerry-washington-abcs-scandal-gets-418091.
66. Sandra Gonzalez, "How the World Became Addicted to *Scandal*," *Entertainment Weekly*, April 5, 2013, http://ew.com/tv/2013/04/05/addicted-scandal/.
67. Aisha Harris, "The Ultimate *Scandal* Live-Tweet Situation Room," *Slate*, November 14, 2013, http://www.slate.com/articles/arts/how_we_watch_tv/2013/11/scandal_on_twitter_all_the_best_live_tweets_including_kerry_washington_s.html.
68. @Larakate, "A Scandal-ous Night on Twitter," *Twitter Blog*, November 14, 2013, https://blog.twitter.com/official/en_us/a/2013/a-scandal-ous-night-on-twitter.html.
69. Elizabeth Wellington, "Social Media Pick Up on *Scandal*," *Philly.com*, April 25, 2013, http://www.philly.com/philly/living/20130425_Social_media_pick_up_on__Scandal.

70. Gonzalez, "How the World Became Addicted to *Scandal*."

71. Tony Maglio, "'Scandal' Shocker: The Shot Heard Round the Net, By the Numbers," *The Wrap*, March 21, 2014, https://www.thewrap.com/scandal-shocker-shot-heard-round-networks-numbers/; "Twitter TV Ratings: Top 10 Series of Fall 2013," *Los Angeles Times*, n.d., http://www.latimes.com/entertainment/envelope/cotown/la-et-tv-twitter-top-20-2013-series-pictures-photogallery.html; Brian Anthony Hernandez, "Top 10 TV Shows on Twitter in the Past Year," *Mashable*, June 4, 2014, http://mashable.com/2014/06/04/top-10-tv-shows-twitter-nielsen-infographic/.

72. Sara Bibel, "ABC Wins Thursday with 'Scandal' Posting Best-Ever Season Finale Ratings," *TV by the Numbers*, April 18, 2014, http://tvbythenumbers.zap2it.com/2014/04/18/abc-wins-thursday-with-scandal-posting-best-ever-season-finale-ratings/255415/.

73. Meredith Blake, "2013 TV Upfronts: ABC Emphasizes 'Scandal' Buzz Over Ratings," *Los Angeles Times*, May 14, 2013, http://articles.latimes.com/2013/may/14/entertainment/la-et-st-abc-upfront-scandal-kerry-washington-brand-20130514.

74. Meghan Casserly, "How 'Scandal's' Shonda Rhimes Became Disney's Primetime Savior," *Forbes*, May 8, 2013, https://www.forbes.com/sites/meghancasserly/2013/05/08/how-scandals-shonda-rhimes-became-disneys-primetime-savior/#4255a1667d5a.

75. "New Nielsen Research Indicates Two-Way Casual Influence Between Twitter Activity and TV Viewership," *Nielsen Press Room*, August 6, 2013, http://www.nielsen.com/us/en/press-room/2013/new-nielsen-research-indicates-two-way-causal-influence-between-.html.

76. Vega, "A Show Makes Friends and History."

77. Gonzalez, "How the World Became Addicted to *Scandal*."

78. Joseph, "Strategically Ambiguous Shonda Rhimes," 311.

79. Everett, "Scandalicious," 38.

80. Warner, "If Loving Olitz Is Wrong, I Don't Want to Be Right: ABC's *Scandal* and the Affect of Black Female Desire," *Black Scholar* 45, no. 1 (2015): 17.

81. Joseph, "Strategically Ambiguous Shonda Rhimes," 311; Gene Demby, "'Scandal': Preposterous, Unmissible, Important," National Public Radio, *Code Switch*, May 18, 2013, https://www.npr.org/sections/codeswitch/2013/05/18/184859810/scandal-preposterous-unmissable-important.

82. Demby, "'Scandal': Preposterous, Unmissible, Important."

83. Choire Sicha, "What Were Black People Talking About on Twitter Last Night?" *The Awl*, November 11, 2009, https://medium.com/the-awl/what-were-black-people-talking-about-on-twitter-last-night-4408ca0ba3d6.

84. Farhad Manjoo, "How Black People Use Twitter," *Slate*, August 10, 2010, http://www.slate.com/articles/technology/2010/08/how_black_people_use_twitter.html.

85. Kimberly C. Ellis, "Why 'They' Don't Understand What Black People Do on Twitter," *Dr. Goddess Blog*, August 2010, http://drgoddess.com/2010/08/why-they-dont-understand-what-black-people-do-on-twitter/; Shani O. Hilton, "You Can Tweet Like This, Or You Can Tweet Like That, Or You Can Tweet Like Us," *PostBourgie*, August 12, 2010, http://|www.postbourgie.com/2010/08/12/you-can-tweet-like-this-or-you-can-tweet-like-that-or-you-can-tweet-like-us/.

86. André Brock Jr., *Distributed Blackness: African American Cybercultures* (New York: NYU Press, 2020), 94–95.

87. Kiana Fitzgerald, "Preview—The Bombastic Brilliance of Black Twitter," *SXTX State*, February 26, 2012, http://sxtxstate.com/2012/02/preview-bombastic-brilliance-black-twitter/; Suzanne Choney and Helen A. S. Popkin, "It's A Black Twitterverse, White People Only Live in It," *Today*, March 10, 2012, https://www.today.com/tech/its-black-twitterverse-white-people-only-live-it-394051.

88. Zizi Papacharissi, "Without You, I'm Nothing: Performances of the Self on Twitter," *International Journal of Communication* no. 6 (2012): 1989–2006; Sarah Florini, "Tweets, Tweeps, and Signifyin': Communication and Cultural Performance on 'Black Twitter,'" *Television & New Media* 15, no. 3 (2014): 223–37; Brock, *Distributed Blackness*.

89. Florini, "Tweets, Tweeps, and Signifyin,'" 225.

90. Brock, *Distributed Blackness*, 108–13.

91. Aaron Smith and Joanna Brenner, "Twitter Use 2012," Pew Research Center, May 31, 2012, http://www.pewinternet.org/2012/05/31/twitter-use-2012/.

92. Brock, *Distributed Blackness*, 214.

93. Mel Stanfill, *Exploiting Fandom: How the Media Industry Seeks to Manipulate Fans* (Iowa City: University of Iowa Press, 2019), 18.

94. Moore, "Shonda Rhimes Lays Claim to Thursday Nights on ABC," Associated Press, September 22, 2014, https://apnews.com/bcbcab2541bf4e50ad7ea709de767de5/shonda-rhimes-lays-claim-thursday-nights-abc.

95. Yvonne Villarreal, "For Shonda Rhimes, a TV Empire Built on High Drama," *Los Angeles Times*, September 12, 2014, http://www.latimes.com/entertainment/tv/sneaks/la-et-st-tv-preview-shonda-rhimes-20140914-story.html.

96. Moore, "Shonda Rhimes Lays Claim to Thursday Nights."

97. Villarreal, "For Shonda Rhimes, a TV Empire Built on High Drama."

98. Gonzalez, "How the World Became Addicted to *Scandal*."

99. Maglio, "Shonda Rhimes on 'How to Get Away with Murder' Title: 'We Don't Consider a Hashtag When We're Writing a Show,'" *The Wrap*, July 16, 2014, https://www.thewrap.com/shonda-rhimes-how-to-get-away-with-murder-abc-hashtag/.

100. Weinstein, "How 'Scandal' Paved the Way."

101. For instance, see "*Monday Night RAW* Tops Nielsen Twitter TV Ratings for the Week of Dec. 22–28," *TV by the Numbers*, December 30, 2014, http://tvbythenumbers.zap2it.com/internet/monday-night-raw-tops-nielsen-twitter-tv-ratings-for-the-week-of-dec-22-28/; "Building Time-Shifted Audiences: Does Social TV Play a Role?" *Nielsen Newswire*, November 4, 2014, http://www.nielsen.com/us/en/insights/news/2014/building-time-shifted-audiences-does-social-tv-play-a-role.html; and "Living Social: How Second Screens Are Helping Making TV Fans," *Nielsen Newswire*, August 4, 2014, http://www.nielsen.com/us/en/insights/news/2014/living-social-how-second-screens-are-helping-tv-make-fans.html.

102. Ellis, "Scheduling," 35.

103. Brian Lowry, "'TGIF'? Well, ABC's Not So Sure Anymore," *Los Angeles Times*, April 14, 2000, http://articles.latimes.com/2000/apr/14/entertainment/ca-19340.

104. See for instance, this clip of *Perfect Strangers* characters Larry and Balki hosting TGIF on November 24, 1989, posted to YouTube, https://www.youtube.com/watch?v=LoTcDkqsiQA.

105. Jonathan Gray, *Show Sold Separately: Promos, Spoilers, and Other Media Paratexts* (New York: NYU Press, 2010), 79.

106. Barker, "'Social TV.'"

107. For instance, see "#BuzzFeedBrews with Ellen Pompeo, Kerry Washington, and Viola Davis," *BuzzFeed*, September 11, 2014, https://livestream.com/BuzzFeedEvents/ShondalandBrews/videos/61703892; Melissa Maerz, "It's Shonda Rhimes's World, We Just Watch It," *Entertainment Weekly*, September 3, 2015, http://ew.com/article/2015/09/03/this-weeks-cover-shondaland-shonda-rhimes-ew/.

108. Karoline Andrea Ihlebaek, Trine Syvertsen, and Espen Ytreberg, "Keeping Them and Moving Them: TV Scheduling in the Phase of Channel and Platform Proliferation," *Television & New Media* 15, no. 5 (2014): 470–75.

109. For more on the growth of connected universes, see Marta Boni, ed., *World Building: Transmedia, Fans, Industries* (Amsterdam: Amsterdam University Press, 2017); Benjamin W. L. Derhy Kurtz and Mélanie Bourdaa, eds., *The Rise of Transtexts: Challenges and Opportunities* (New York: Routledge, 2016).

110. Gillan, *Television Brandcasting*.

111. Rick Kissell, "Killer Start Thursday for ABC's 'How to Get Away with Murder,'" *Variety*, September 26, 2014, http://variety.com/2014/data/ratings/killer-start-thursday-for-abcs-how-to-get-away-with-murder-1201314466/.

112. "Weekly Top Ten Series, September 22–September 28," *Nielsen Social*, September 30, 2014, http://www.nielsensocial.com/rating/week-of-09-22-14/.

113. Lisa de Moraes, "Full 2014–2015 TV Series Rankings: Football & 'Empire' Ruled," *Deadline*, May 21, 2015, http://deadline.com/2015/05/2014-15-full-tv-season-ratings-shows-rankings-1201431167/; "The Making of Social TV: Loyal Fans and Big Moments Build Program-Related Buzz," *Nielsen Newswire*, August 24, 2015, http://www.nielsen.com/us/en/insights/news/2015/the-making-of-social-tv-loyal-fans-and-big-moments-build-program-buzz.html.

114. Saba Hamedy, "Tweeting as If It's Thursday," *Los Angeles Times*, November 25, 2014, https://www.latimes.com/entertainment/envelope/cotown/la-et-ct-tgit-abc-20141126-story.html.

115. Hamedy, "Tweeting as If It's Thursday."

116. Hamedy, "Tweeting as If It's Thursday."

117. Hamedy, "Tweeting as If It's Thursday."

118. Mary Ingram-Waters and Leslie Balderas, "Blurring Production Boundaries within Fan Empowerment," in *Adventures in Shondaland: Identity Politics and the Power of Representation*, eds. Michaela D. E. Meyer and Rachel Alicia Griffin (New Brunswick: Rutgers University Press, 2018), 197–213.

119. Natalie Abrams, "*Scandal*: Why ABC Decided to End the Series," *Entertainment Weekly*, May 16, 2017, http://ew.com/tv/2017/05/16/scandal-season-7-end/.

120. Nellie Andreeva, "'Scandal'/'How to Get Away with Murder' Crossover: Get Details on Big Shondaland Event From 'HTGAWM' Creator—Q&A," *Deadline*, January 3, 2018, http://deadline.com/2018/01/scandal-how-to-get-away-with-murder-crossover-shonda-rhimes-kerry-washington-viola-davis-htgawm-creator-qa-1202234080/.

121. Danielle Turchiano, "Shonda Rhimes Partners with Hearst, Launches New Lifestyle Website," *Variety*, September 18, 2017, http://variety.com/2017/tv/news/shonda-rhimes-launches-lifestyle-website-1202561585/.

122. Andrew Wallenstein, "Netflix Lures Shonda Rhimes Away from ABC Studios," *Variety*, August 13, 2017, http://variety.com/2017/digital/news/netflix-lures-shonda-rhimes-away-from-abc-studios-report-1202526464/.

CHAPTER 2

1. Oliver Gettell, "*Preacher* Pilot with Directors Commentary Released Online," *Entertainment Weekly*, May 26, 2016, https://ew.com/article/2016/05/26/preacher-pilot-directors-commentary-seth-rogen/.

2. "AMC Announces Story Sync," *Daily Dead*, February 7, 2012, https://dailydead.com/amc-announces-the-walking-dead-story-sync/.

3. Jason Mittell, "Forensic Fandom and the Drillable Text," *Spreadable Media*, 2013, http://spreadablemedia.org/essays/mittell/.

4. Suzanne Scott, *Fake Geek Girls: Fandom, Gender, and the Convergence Culture Industry* (New York: NYU Press, 2019), 77.

5. Taylor Nygaard and Jorie Lagerwey, "Broadcasting Quality: Re-Centering Feminist Discourse with *The Good Wife*," *Television & New Media* 18, no. 2 (2017): 108–110.

6. Derek Johnson, "Devaluing and Revaluing Seriality: The Gendered Discourses of Media Franchising," *Media, Culture, and Society* 33, no. 7 (2011): 1080.

7. John T. Caldwell, "Convergence Television: Aggregating Form and Repurposing Content in the Culture of Conglomeration," in *Television After TV: Essays on a Medium in Transition*, eds. Spigel and Jan Olsson (Durham: Duke University Press, 2004), 41–74; Barbara Klinger, *Beyond the Multiplex: Cinema, New Technologies, and the Home* (Berkley: University of California Press, 2006), 7–8.

8. "The Walking Dead Story Sync," Emmys.com, n.d., http://www.emmys.com/shows/walking-dead-story-sync.

9. Bryan Bishop, "How a Second-Screen App Made 'The Walking Dead' Come Alive," *The Verge*, February 13, 2014, http://www.theverge.com/entertainment/2014/2/13/5406498/how-a-second-screen-app-made-the-walking-dead-come-alive.

10. Megan Daley, "*The Walking Dead* is Twitter's Biggest Show," *Entertainment Weekly*, June 1, 2015, http://ew.com/article/2015/06/01/walking-dead-twitter/.

11. Since fall 2017, the Story Sync landing page for *The Walking Dead* notes, "Story Sync will not run during new episodes of *The Walking Dead*. But you can still browse past syncs for Season 4–7, and check out all-new apps & games via the link below." "*The Walking Dead* Story Sync," *AMC*, n.d., http://www.amc.com/shows/the-walking-dead/story-sync/.

12. Paul Grainge, "Introduction: Ephemeral Media," in *Ephemeral Media: Transitory Screen Culture from Television to YouTube*, ed. Paul Grainge (London: British Film Institute, 2011), 2.

13. Amelie Hastie, "Detritus and the Moving Image: Ephemera, Materiality, History," *Journal of Visual Culture* 6, no. 2 (2007): 171–74.

14. Klinger, *Beyond the Multiplex*, 200.

15. Gerard Genette, *Paratexts: The Thresholds of Interpretation*, trans. Jane E. Lewin (Cambridge: Cambridge University Press, 1997).

16. Jonathan Gray, *Show Sold Separately: Promos, Spoilers, and Other Media Paratexts* (New York: NYU Press, 2010), 6.

17. Gray, *Show Sold Separately*, 26.

18. Mikhail Iampolski, *The Memory of Tiresias: Intertextuality and Film*, trans. Harsha Ram (Berkeley: University of California Press, 1998), 2.

19. Michael Riffaterre, "Compulsory Reader Response: The Intertextual Drive," in *Intertextuality: Theories and Practices*, eds. Michael Worton and Judith Still (Manchester: Manchester University Press, 1990), 76.

20. Robert Brookey and Jonathan Gray, "Not Merely Para: Continuing Steps in Paratextual Research," *Critical Studies in Media Communication* 34, no. 2 (2017): 108–10.

21. Melissa Aronczyk, "Portal or Police? The Limits of Promotional Paratexts," *Critical Studies in Media Communication* 34, no. 2 (2017): 115.

22. Hye Jin Lee and Mark Andrejevic, "Second-Screen Theory: From the Democratic Surround to the Digital Enclosure," in *Connected Viewing: Selling, Streaming, & Sharing Media in the Digital Era*, eds. Jennifer Holt and Kevin Sanson (New York: Routledge, 2014); Julia Kristeva, *Desire in Language: A Semiotic Approach to Literature and Art*, trans. Thomas Gora et al., ed. Leon Roudiez (Oxford: Blackwell, 1980), 65.

23. Chris Foresman, "iPad Adoption Rate Fastest in Electronics Product History," CNN, October 5, 2010, http://www.cnn.com/2010/TECH/gaming.gadgets/10/05/ipad.adoption.ars/.

24. Paul McDonald, *Video and DVD Industries* (London: British Film Institute, 2008).

25. Klinger, *Beyond the Multiplex*, 61.

26. Mark Parker and Deborah Parker, *The DVD and the Study of Film: The Attainable Text* (New York: Palgrave Macmillan, 2011), xvii.

27. Parker and Parker, *The DVD and the Study of Film*, 92.

28. McDonald, *Video and DVD Industries*, 6.

29. Matt Hills, "From the Box in the Corner to the Box Set on the Shelf: 'TVIII' and the Cultural/Textual Valorizations of DVD," *New Review of Film and Television Studies* 5, no. 1 (2007): 45; Derek Kompare, "Publishing Flow: DVD Box Sets and the Reconception of Television," *Television & New Media* 7, no. 4 (2006): 336.

30. Klinger, *Beyond the Multiplex*, 71.

31. Robert Alan Brookey and Robert Westerfelhaus, "Hiding Homoeroticism in Plain View: The *Fight Club* DVD as Digital Closet," *Critical Studies in Media Communication* 19, no. 1 (2002): 23.

32. James Walters, "Repeat Viewings: Television Analysis in the DVD Age," in *Film and Television After DVD*, eds. James Bennett and Tom Brown (London: Routledge, 2008), 70–71; Craig Hight, "Making-of Documentaries on DVD: *The Lord of the Rings* Trilogy and Special Editions," *Velvet Light Trap* 56 (2005): 14.

33. Greg M. Smith, *On A Silver Platter: CD-ROMs and The Promises of a New Technology* (New York: NYU Press, 1999), 10–14.

34. Caldwell, "Prefiguring DVD Bonus Tracks: Making-ofs and Behind-the-Scenes as Historic Television Programming Strategies Prototypes," in *Film and Television After DVD*, eds. James Bennett and Tom Brown (London: Routledge, 2008), 150.

35. Brookey and Westerfelhaus, "Hiding Homoeroticism," 25.

36. Caldwell, "'Second-Shift' Media Aesthetics: Programming, Interactivity, and User Flows," in *New Media: Theories and Practices of Digitextuality*, eds. John T. Caldwell and Anna Everett (London: Routledge, 2003), 136.

37. Hills, "From the Box in the Corner," 53.

38. Klinger, *Beyond the Multiplex*, 56; Parker and Parker, *The DVD and the Study of Film*, xii.

39. McDonald, *Video and DVD Industries*, 7.

40. Andrew Wallenstein, "Home Entertainment 2016 Figures: Streaming Eclipses Disc Sales for the First Time," *Variety*, January 6, 2017, http://variety.com/2017/digital/news/home-entertainment-2016-figures-streaming-eclipses-disc-sales-for-the-first-time-1201954154/.

41. Ryan Faughnder, "Home Video Sales Shrank Again in 2016 as Americans Switched to Streaming," *Los Angeles Times*, January 6, 2017, http://www.latimes.com/business/hollywood/la-fi-ct-home-video-decline-20170106-story.html.

42. For instance, see Matt Singer, "Vanishing DVD Extras on Netflix," *IFC*, December 10, 2010, https://www.ifc.com/2010/12/netflix-and-dvd-extras; "More Studios Stripping Special Features from More DVDs?" *Hacking Netflix*, April 14, 2010, http://www.hackingnetflix.com/2010/04/studios-stripping-special-features-from-more-dvds.html.

43. John Tomlinson, *The Culture of Speed: The Coming of Immediacy* (London: Sage, 2007), 74.

44. Joshua Green, "Why Do They Call It TV When It's Not on The Box? 'New' Television Services and Old Television Functions," *Media International Australia*, no. 126 (2008): 96.

45. Max Dawson, "Little Players, Big Shows: Format, Narrative, and Style on Television's New Smaller Screens," *Convergence* 13, no. 3 (2007): 231–50.

46. Gillian Doyle, "From Television to Multi-Platform: Less from More or More for Less?" *Convergence* 16, no. 4 (2010): 431–49.

47. Henry Jenkins, *Convergence Culture: When Old and New Media Collide* (New York: NYU Press, 2006), 95–96.

48. Gray, *Show Sold Separately*, 210–15. See also Henry Jenkins, "'We Had So Much Story to Tell': *The Heroes* Comics as Transmedia Storytelling," *Confessions of An Aca-Fan*, December 3, 2007, http://henryjenkins.org/blog/2007/12/we_had_so_many_stories_to_tell.html.

49. For detailed accounts of *The Lost Experience* and fan response, see its dedicated page on *Lostpedia*, noted fan wiki for *Lost*: http://lostpedia.wikia.com/wiki/The_Lost_Experience.

50. Jeffrey Sconce, "What If? Charting Television's New Textual Boundaries," in *Television After TV: Essays on a Medium in Transition*, eds. Lynn Spigel and Jan Olsson (Durham: Duke University Press, 2004), 95.

51. Mittell, *Complex TV: The Poetics of Contemporary Television Storytelling* (New York: NYU Press, 2015), 306–310.

52. Dawson, "Television Abridged: Ephemeral Texts, Monumental Seriality, and TV-Digital Media Convergence," in *Ephemeral Media: Transitory Screen Culture from Television to YouTube*, ed. Paul Grainge (London: British Film Institute, 2011), 38–40; Grainge, "Ephemeral Media," 10; Gray, *Shows Sold Separately*, 23–24.

53. Dawson, "Television Abridged," 46–50.

54. Green, "Why Do They Call It TV," 97.

55. Kim Bjarkman, "To Have and to Hold: The Video Collector's Relationship to an Ethereal Medium," *Television & New Media* 5, no. 3 (2004): 230.

56. Ryan Lawler, "Viacom Joins Comcast, NBCU, and HBO As Social TV Start-up Zeebox's Latest Strategic Partner," *TechCrunch*, October 4, 2012, https://techcrunch.com/2012/10/04/zeebox-viacom/.

57. Richard Lawler, "Yahoo Buys TV Companion App Developer IntoNow and Its Database of Sounds," *Engadget*, April 25, 2011, https://www.engadget.com/2011/04/25/yahoo-buys-tv-companion-app-developer-intonow-and-its-database-o/; Richard Lawler, "Yahoo Connected TV Setups Draw Web TV Closer in 2012 with Mobile Apps, IntoNow," *Engadget*, January 12, 2012, https://www.engadget.com/2011/04/25/yahoo-buys-tv-companion-app-developer-intonow-and-its-database-o/.

58. Ryan Lawler, "Zeebox Partners with Comcast, NBC Universal, & HBO to Launch Companion App in the US," *TechCrunch*, September 26, 2012, https://techcrunch.com/2012/09/26/zeebox-us-launch-comcast-nbcu-hbo/.

59. Jennifer Van Grove, "USA Network Builds Second Screen Experience Around 'Psych,'" *Mashable*, December 15, 2010, https://mashable.com/2010/12/15/usa-psych-vision/#bHBsTixImqqj.

60. Andrew Wallenstein, "'X Factor' Expands Digital Domain," *Variety*, November 11, 2011, http://www.variety.com/article/VR1118046251%27X.

61. Cory Bergman, "MLB Hits 3 Million App Downloads 8 Days into Season," *Lost Remote*, April 12, 2012, http://www.adweek.com/lostremote/mlb-hits-3-million-app-downloads-in-just-8-days/28786.

62. Richard Lawler, "NBC's 2012 London Olympics Second Screen and Streaming Apps for Android and iOS Launch Today," *Engadget*, July 12, 2012, https://www.engadget.com/2012/07/12/nbc-2012-london-olympics-second-screen-streaming-apps/.

63. Natan Edelsburg, "*Breaking Bad*'s Cast and AMC Use Social Media with Fans," *Lost Remote*, July 15, 2012, http://www.adweek.com/lostremote/how-breaking-bads-cast-and-amc-use-social-with-fans-video/32996.

64. Liz Shannon Miller, "Can *Breaking Bad*'s Story Sync Get Viewers to Give Up Their DVRs?" *Gigaom*, September 9, 2012, https://gigaom.com/2012/09/09/can-breaking-bads-story-sync-get-viewers-to-give-up-their-dvrs/.

65. "Story Sync Turns *Breaking Bad* into Two-Screen Experience with Polls and Video," *AMC.com*, n.d., http://www.amc.com/shows/breaking-bad/talk/2012/07/breaking-bad-story-sync.

66. Hare, "Twice as Much TV?"

67. Elizabeth Evans, *Transmedia Television: Audiences, New Media, and Daily Life* (New York: Routledge, 2011), 32.

68. Evans, *Transmedia Television*, 33–34.

69. Hare, "Twice as Much TV?"
70. Klinger, *Beyond the Multiplex*, 72.
71. Caldwell, *Production Culture*, 284.
72. Caldwell, "Prefiguring DVD Bonus Tracks," 141.
73. Klinger, *Beyond the Multiplex*, 72.
74. Parker and Parker, *The DVD and the Study of Film*, 25.
75. Brookey and Westerfelhaus, "Hiding Homoeroticism," 23.
76. Mittell, *Complex TV*, 261–62.
77. Sconce, "What If?" 95.
78. Will Brooker, "Living on *Dawson's Creek*: Teen Viewers, Cultural Convergence, and Television Overflow," *International Journal of Cultural Studies* 4, no. 4 (2001): 456–72.
79. Klinger, *Beyond the Multiplex*, 73.
80. Brooker, "Television Out of Time: Watching Cult Shows on Download," in *Reading Lost: Perspectives on a Hit Television Show*, ed. Roberta Pearson (London: I. B. Tauris, 2009), 51–72.
81. Caldwell, *Production Culture*, 291, 298.
82. Jenkins, *Convergence Culture*, 133.
83. Henry Jenkins and Nico Carpentier, "Theorizing Participatory Intensities: A Conversation about Participation and Politics," *Convergence* 19, no. 3 (2013): 272.
84. Jo T. Smith, "DVD Technologies and the Art of Control," in *Film and Television After DVD*, eds. James Bennett and Tom Brown (London: Routledge, 2008), 141–42.
85. Bishop, "How a Second-Screen App Made 'The Walking Dead' Come Alive."
86. Green, "Why Do They Call It TV?" 97.
87. Caldwell, *Production Culture*, 300.
88. David Carr, "At AMC, Zombies Topple Network TV," *New York Times*, March 3, 2013, http://www.nytimes.com/2013/03/04/business/media/walking-dead-helps-solidify-amcs-ratings-success.html.
89. Malcolm Moore, "Film Studios Select Movie-Fan Data for Starring Role to Drive Ticket Sales," *Financial Times*, September 7, 2016, https://www.ft.com/content/8489fd08-15bc-11e6-b197-a4af20d5575e.
90. Joanna Robinson, "*Breaking Bad* Alum Giancarlo Esposito Explains the Return of Gus Fring on *Better Call Saul*," *Vanity Fair*, January 14, 2017, https://www.vanityfair.com/hollywood/2017/01/gus-fring-better-call-saul-season-3-giancarlo-esposito.
91. Robinson, "How Online Fandom Is Shaping TV in 2017," *Vanity Fair*, January 19, 2017, https://www.vanityfair.com/hollywood/2017/01/online-fandom-tv-2017-riverdale-twin-peaks-better-call-saul-time-after-time.
92. LeCareJames, "A Great Catch You May Have Missed If You Don't Use Story Sync," Reddit thread, 2017, https://www.reddit.com/r/betterCallSaul/comments/4eebjh/a_great_catch_you_might_have_missed_if_you_dont/.
93. Mittell, *Complex TV*, 288–89.
94. AllureFX, "What Is the Likeliest Outcome for Walt (Previous Story Sync)?" Reddit thread, 2013, https://www.reddit.com/r/breakingbad/comments/1m2zj3/what_is_the_likeliest_outcome_for_walt_previous/.

95. CapnJackson, "[SPOILERS] Story Sync Dropped the Ball?" Reddit thread, 2016, https://www.reddit.com/r/thewalkingdead/comments/4d96ht/spoilers_story_sync_dropped_the_ball/.

96. WhatTheMess, "[Spoilers] Subtle Differences in These Pages from Tonight's Story Sync," Reddit thread, 2016, https://www.reddit.com/r/thewalkingdead/comments/3uswyl/spoilers_subtle_difference_in_these_pages_from/.

97. Roninjinn, "Possible [SPOILERS] from FTWD Story Sync Last Night, Origins of the Outbreak?" Reddit thread, 2015, https://www.reddit.com/r/thewalkingdead/comments/3i76rs/possible_spoilers_from_ftwd_story_sync_last_night/.

98. Grainge, "Ephemeral Media," 9.

99. Todd Spangler, "Showtime and LG Think They've Cracked the Code on Interactive TV," *Variety*, August 15, 2013, http://variety.com/2013/digital/news/showtime-and-lg-think-theyve-cracked-code-on-interactive-tv-1200578659/.

100. KP, "HBO Launches Interactive Viewing Experience for *Game of Thrones*," *Two Cents TV*, March 29, 2012, http://twocentstv.com/hbo-launches-interactive-viewing-experience-for-game-of-thrones/.

101. Christina Warren, "HBO Gets Social with HBO Connect," CNN, May 25, 2011, http://www.cnn.com/2011/TECH/social.media/05/25/hbo.connect.mashable/.

102. Spangler, "TV Viewers Aren't Thrilled with Second Screen Synchronized Content, Study Finds," *Variety*, January 9, 2014, http://variety.com/2014/digital/news/tv-viewers-arent-thrilled-with-second-screen-synchronized-content-study-finds-1201040757/.

103. "HBO Reconsiders Second Screening," *Warc*, July 29, 2015, http://www.warc.com/LatestNews/News/EmailNews.news?ID=35146.

104. Jeff Baumgartner, "Showtime to Discontinue 'SHO Sync,'" *Multichannel News*, July 9, 2015, http://www.multichannel.com/news/next-tv/showtime-discontinue-sho-sync/392052.

105. See the *Game of Thrones* Viewers Guide: viewers-guide.hbo.com/gameofthrones.

106. *Game of Thrones* Viewers Guide.

107. Immallama, "I Really Missed Story Sync This Season," Reddit thread, 2017, https://www.reddit.com/r/betterCallSaul/comments/6ibpe3/i_really_missed_story_sync_this_season/?st=jdhq4nxt&sh=6cc6e7bb; About60zombies, "Are They Still Doing Story Sync for Season 3?" Reddit thread, 2017, https://www.reddit.com/r/FearTheWalkingDead/comments/71wrkz/are_they_still_doing_story_sync_for_season_3/?st=jdhq5o5n&sh=6611245d.

CHAPTER 3

1. "Self-reported positioning" was the more technical term to describe what would become the check-in. See Wolfgang Broll et al., "Meeting Technology Challenges of Pervasive Augmented Reality Games," Presented at the *Proceedings of the 5th ACM SIGCOMM Workshop on Network and Support System for Games—NetGames '06*, October 30–31, 2006, article 28, n.p.

2. M. G. Siegler, "HBO Sinks Its Teeth into GetGlue To Reward Fans for Checking In to Hit Shows," *TechCrunch*, July 28, 2010, https://techcrunch.com/2010/07/28/getglue-hbo/; Tom Cheredar, "Miso Transforms *Dexter*'s Season Premiere into a 'Pop-Up Video' Experience," *Venture Beat*, September 27, 2011, https://venturebeat.com/2011/09/27/miso-dexter/.

3. "How Does Viggle Know What I'm Watching," *Viggle*, accessed December 10, 2017, http://functionxinc.desk.com/customer/portal/articles/326522-how-does-viggle-know-what-i-m-watching; dead link.

4. For a survey of check-in demographics, see Mike Proulx and Stacey Shepatin, *Social TV: How Marketers Can Research and Engage Audiences by Connecting Television to the Web, Social Media, and Mobile* (Hoboken: Wiley, 2012), 57–80. See also, Danielle Johnsen Karr, "GetGlue Grows User Base to 2MM, Unlocking Opportunities for Brands," *360i*, January 13, 2012, http://blog.360i.com/social-marketing/get-glue. For a broader report on Social TV usage, see "Report: U.S. Media Trends by Demographic," *Nielsen*, April 27, 2012, http://www.nielsen.com/us/en/insights/news/2012/report-u-s-media-trends-by-demographic.html.

5. "FAQ," *GetGlue*, accessed February 27, 2018, http://getglue.com/faq; dead link.

6. "How Does Viggle Know What I'm Watching," *Viggle*, accessed December 10, 2017, http://functionxinc.desk.com/customer/portal/articles/326522-how-does-viggle-know-what-i-m-watching; dead link.

7. "Rewards," *Viggle* (iPhone platform), accessed December 3, 2012.

8. Todd Spangler, "GetGlue Thinks It Has Fighting Chance Against Twitter, Facebook in Social TV," *Variety*, September 5, 2013, http://variety.com/2013/digital/news/getglue-thinks-it-has-fighting-chance-against-twitter-facebook-in-social-tv-1200599891/; Jeff Baumgartner, "Viggle Extends Second Screen Ad Bridge," *Multichannel News*, August 12, 2013, http://www.multichannel.com/technology/viggle-extends-second-screen-ad-bridge/144906; Natan Edelsbrun, "The NBA and Viggle Launch Second Screen Partnership for the 2013 NBA Playoffs," *Lost Remote*, May 13, 2013, http://lostremote.com/the-nba-and-viggle-launch-second-screen-partnership-for-the-2013-nba-playoffs_b37580; Lisa Lacy, "Miso Launches Social TV Platform Quips," *Clickz*, November 12, 2012, http://www.clickz.com/clickz/news/2224263/miso-launches-social-tv-platform-quips.

9. Eliot Van Buskirk, "Comcast Unveils 'Tunerfish' Social TV App," *Wired*, May 24, 2010, https://www.wired.com/2010/05/comcast-unveils-tunerfish-social-tv-app/; Tim Conneally, "CBS Interactive Launches the Foursquare of TV Viewing, TV.com Relay," *Betanews*, August 12, 2010, https://betanews.com/2010/08/12/cbs-interactive-launches-the-foursquare-of-tv-viewing-tv-com-relay/; Pete Pachal, "Is This the Second-Screen TV App That Finally Goes Mainstream?" *Mashable*, September 27, 2012, http://mashable.com/2012/09/27/zeebox/; and Ben Drawbaugh, "NextGuide for iPad Intends to Get You to Stop Using the Grid Guide," *Engadget*, September 7, 2012, https://www.engadget.com/2012/09/07/nextguide-for-ipad-intends-to-get-you-to-stop-using-the-grid-gui/.

10. Ryan Lawler, "Social TV Start-up Dijit Buys Miso to Re-Define the Second Screen," *TechCrunch*, February 1, 2013, https://techcrunch.com/2013/02/01/dijit-buys-miso-really-yes-really/.

11. Simon Dumenco, "i.TV is Buying GetGlue; Here's What That Means for Social TV," *Ad Age*, November 6, 2013, http://adage.com/article/the-media-guy/i-tv-buying-getglue-means-social-tv/245137/; Cory Bergman, "What the Viggle-GetGlue Acquisition Means for the Battle over the Second Screen," *Lost Remote*, November 19, 2012, http://lostremote.com/what-the-viggle-getglue-acquisition-means-for-the-battle-over-the-second-screen_b35220; Catherine Shu, "GetGlue-Viggle Merger Called Off," *TechCrunch*, January 13, 2013, http://techcrunch.com/2013/01/13/736122/; Jon Fingas, "i.TV Acquires GetGlue to Boost Its Stake in Second Screen Viewing," *Engadget*, November 6, 2013, https://www.engadget.com/2013/11/06/i-tv-acquires-getglue/.

12. Janko Roettgers, "GetGlue Successor TVtag is Shutting Down," *Gigaom*, December 19, 2014, https://gigaom.com/2014/12/19/getglue-successor-tvtag-is-shutting-down/.

13. Spangler, "Viggle Buys Dijit Media, in Another Second-Screen TV Mashup," *Variety*, January 29, 2014, http://variety.com/2014/digital/news/viggle-buys-dijit-media-in-another-second-screen-tv-mashup-1201076325/; "Perk.com Inc. Announces Closing of Acquisition of Viggle App & Assets," *Business Wire*, February 8, 2016, http://www.businesswire.com/news/home/20160208005554/en/Perk.com%C2%A0Inc.-Announces-Closing-Acquisition-Viggle-App-Assets.

14. Ryan M. Milner, "Working for the Text: Fan Labor and the New Organization," *International Journal of Cultural Studies* 12, no. 5 (2009): 495; Howard Rheingold, *The Virtual Community: Homesteading on the Electronic Frontier* (Malden: MIT Press, 1993).

15. Lewis Hyde, *The Gift: Creativity and the Artist in the Modern World*, 25th Anniversary Edition (New York: Vintage Books, 2009), 100–105.

16. Abigail De Kosnik, "Should Fan Fiction Be Free?" *Cinema Journal* 48, no. 4 (2009): 118–24; Xiaochang Li, "Fanfic, Inc.: Another Look at Fanlib.com," *MIT Convergence Culture Consortium C3 Weekly Update*, December 7, 2007, http://www.convergenceculture.org/htmlnewsletter/weeklyupdate_20071207.html; Karen Hellekson, "A Fannish Field of Value: Online Gift Culture," *Cinema Journal* 48, no. 4 (2009): 113–18.

17. Xiaochang Li, *More than Money Can Buy: Locating Value in Spreadable Media*, (Cambridge: MIT Convergence Culture Consortium, 2009), 24, http://convergenceculture.org/research/C3LocatingValueWhitePaper.pdf.

18. Richard Barbrook, "The Hi-Tech Gift Economy," *First Monday* 3, no. 12 (1998/2005), http://firstmonday.org/htbin/cgiwrap/bin/ojs/index.php/fm/article/view/631/552.

19. Paul Booth, *Digital Fandom: New Media Studies* (New York: Peter Lang, 2010), 24–27.

20. Suzanne Scott, "Repackaging Fan Culture: The Regifting Economy of Ancillary Content Models," *Transformative Works and Cultures*, no. 3 (2009), https://doi.org/10.3983/twc.2009.0150.

21. Michael Hardt, "Affective Labor," *Boundaries* 2 (1999): 89–100.

22. Jenna Wortham, "Foursquare Signs a Deal with Zagat," *New York Times*, February 9, 2010, https://bits.blogs.nytimes.com/2010/02/09/foursquare-inks-a-deal-with-zagat/; Leena K. Rao, "Foursquare Gets Lucky (Magazine) and A Deal with Condé Nast," *TechCrunch*, February 12, 2010, https://techcrunch.com/2010/02/12/foursquare-gets-lucky-magazine/; Wortham, "A Start-Up Matures, Working with AmEx," *New York Times*, June 22, 2011, http://www.nytimes.com/2011/06/23/technology/23locate.html.

23. Allison Mooney, "Beyond the Badge: Big Media Brands Strike Foursquare Deals," *Ad Age*, February 8, 2010, http://adage.com/article/digitalnext/mobile-marketing-big-media-brands-strike-foursquare-deals/141977/.

24. Lydia Leavitt, "Top 7 Location-Based Apps to Use When Foursquare is Down," *TheNextWeb*, October 5, 2010, https://thenextweb.com/socialmedia/2010/10/05/top-7-location-based-apps-to-use-when-foursquare-is-down/#.tnw_N9mk0rF4.

25. Ryan Singel, "Facebook Launches 'Check-In' Service to Connect People in Real Space," *Wired*, August 18, 2010, https://www.wired.com/2010/08/watch-facebooks-location-sharing-announcement-live/.

26. Lou Dubois, "GetGlue: The Foursquare of Entertainment," *Inc.*, October 25, 2010, http://www.inc.com/news/articles/2010/10/getglue-the-foursquare-of-entertainment-checkin.html.

27. Simon Dumenco, "Will Viggle, the 'Loyalty Program for TV' From Robert F.X. Silverman, Transform Television?" *Ad Age*, January 18, 2012, http://adage.com/article/digital/viggle-loyalty-program-tv-transform-television/232169/.

28. Alexander M. C. Halavais, "A Genealogy of Badges: Inherited Meaning and Monstrous Moral Hybrids," *Information, Communication & Society* 15, no. 3 (2012): 354–73.

29. Cameron Lister et al., "Just a Fad? Gamification in Health and Fitness Apps," *JMIR Serious Games* 2, no. 2 (2014): n.p.; Sara Corbett, "Learning by Playing: Video Games in the Classroom," *New York Times Magazine*, September 15, 2010, http://www.nytimes.com/2010/09/19/magazine/19video-t.html; and Shantanu Sinha, "Motivating Students and the Gamification of Learning," *Huffington Post*, February 14, 2012, https://www.huffingtonpost.com/shantanu-sinha/motivating-students-and-t_b_1275441.html.

30. Juho Hamari and Jonna Koivisto, "'Working Out for Likes': An Empirical Study on Social Influence in Exercise Gamification," *Computers in Human Behavior*, no. 50 (2015): 333–47; Alaa AlMarshedi et al., "Gamifying Self-Management of Chronic Illnesses: A Mixed-Methods Study," *JMIR Serious Games* 4, no. 2 (2016): n.p.

31. Ian Bogost, "Persuasive Games: Exploitationware," *Gamasutra*, May 3, 2011, http://www.gamasutra.com/view/feature/134735/persuasive_games_exploitationware.php.

32. Dumenco, "Loyalty Program."

33. John Fiske, "The Cultural Economy of Fandom," in *The Adoring Audience: Fan Culture and Popular Media*, ed. Lisa A. Lewis (New York: Routledge, 1992), 37, 34.

34. Fiske, "The Cultural Economy of Fandom," 37.

35. Cornel Sandvoss, "Fans Online: Affective Media Consumption and Production in the Age of Convergence," in *Online Territories: Globalization, Mediated Practice, and Social Space*, eds. Miyase Christensen, André Jansson, and Christian Christensen (New York: Peter Lang, 2011), 59–60; Suzanne Scott, "'Authorized Resistance': Is Fan Production Frakked?" in *Cylons in America: Critical Studies in Battlestar Galactica*, eds. Tiffany Potter and C. W. Marshall (New York: Continuum, 2008), 212; Scott, "'Who's Steering the Mothership?' The Role of the Fanboy Auteur in Transmedia Storytelling," in *The Participatory Cultures Handbook*, eds. Aaron Delwiche and Jennifer Jacobs Henderson (New York: Routledge, 2013), 46–47.

36. Booth, *Negotiating Fandom and Media in the Digital Age* (Iowa City: University of Iowa Press, 2015), 42.

37. Vincent Miller, "New Media, Networking, and Phatic Culture," *Convergence* 14, no. 4 (2008): 395.

38. Ethan Tussey, *The Procrastination Economy: The Big Business of Downtime* (New York: NYU Press, 2018), 143.

39. Geert Lovink, *Networks Without A Cause: A Critique of Social Media* (Cambridge: Polity Press, 2011), 52.

40. Dubois, "GetGlue: The Foursquare of Entertainment."

41. Spangler, "GetGlue Thinks It Has Fighting Chance."

42. Kunur Patel, "Check in Before You Check out That Show," *Ad Age*, September 6, 2010, http://adage.com/article/media/check-apps-a-win-win-tv-viewers-networks/145736/.

43. Carolin Gerlitz and Anne Helmond, "The Like Economy: Social Buttons and the Data-Intensive Web," *New Media & Society* 15, no. 8 (2013): 1358.

44. June Thomas, "An Oddly Addiction New Social Network for TV Fanatics," *Slate*, March 12, 2012, http://www.slate.com/blogs/browbeat/2012/03/12/getglue_an_oddly_addictive_new_social_network_for_tv_fanatics.single.html.

45. Megan O'Neill, "GetGlue Launches A New Real-Time Conversation Experience," *Adweek*, October 25, 2011, http://www.adweek.com/digital/getglue-conversation/.

46. Tommie, Comment on *The Walking Dead*, GetGlue, November 26, 2012, accessed December 3, 2012, http://getglue.com/conversation/tohkeeoh/2012-11-26T02:52:43Z, dead link.

47. Various, Comments on *The Vampire Diaries*, GetGlue, October 23, 2013, accessed October 31, 2013, dead link.

48. Nunya "Shitagi Nashi" Business, Comment on *Saturday Night Live*, GetGlue, December 8, 2012, accessed December 8, 2012, http://getglue.com/conversation/chicitygurl/2012-12-09T04:46:38Z, dead link.

49. Candibug76, Comment on *Revenge*, GetGlue, November 26, 2012, accessed December 3, 2012, http://getglue.com/conversation/candi_a__anderson/2012-11-26T02:57:02Z, dead link.

50. Alex, "Introducing Guru Special Powers and Giveaways," *GetGlue*, February 2, 2010, accessed December 2, 2012, http://blog.getglue.com/?p=4034.

51. "Full Sticker FAQ," Sticker FAQ, http://stickerfaq.blogspot.com/p/full-sticker-faq.html.

52. Marc, Comment on *New Girl*, Get Glue, October 22, 2013, accessed December 3, 2012, dead link.

53. "*New Girl* Stickers," GetGlue, accessed December 2, 2012, http://o.getglue.com/stickers/groups?group=New+Girl, dead link; "*Ben and Kate* Stickers," *GetGlue*, accessed December 2, 2012, http://o.getglue.com/stickers/groups?group=Ben+and+Kate, dead link.

54. For more examples of limited stickers, see the GetGlue Wiki, http://getglue.wikia.com/wiki/Category:Limited.

55. Cotton Delo, "Gap Sponsors New Partnership between *Entertainment Weekly*, GetGlue," *Ad Age*, September 15, 2011, http://adage.com/article/media/gap-sponsors-deal-getglue-entertainment-weekly/229788/.

56. Sarah Thornton, *Club Cultures: Music, Media, and Subcultural Capital* (Middletown: Wesleyan University Press, 1996), 11–14.

57. Fiske, "The Cultural Economy of Fandom," 34.

58. Chuck Tryon, *On-Demand Culture: Digital Delivery and the Future of Movies* (New Brunswick: Rutgers University Press, 2013), 128.

59. Megan O'Neill, "Top 10 TV Shows in October According to GetGlue Check-Ins," *Adweek*, November 7, 2011, http://www.adweek.com/digital/top-10-tv-shows-in-october/.

60. Proulx and Shepatin, *Social TV*, 64.

61. Warner Bros. TV, "Fans of Warner Bros. Television's *The Big Bang Theory* Check in More than Two Million Times on GetGlue.com, Ranking as the #1 Series on the Site and Doubling in Popularity in the Past Three Months," *PR Web*, March 8, 2012, http://www.prweb.com/releases/the_big_bang_theory/wb_getglue/prweb9266222.htm.

62. Ryan Lawler, "With 3 Million Users, GetGlue Goes Big with a New Social TV App Built Just for the iPad," *TechCrunch*, August 16, 2012, https://techcrunch.com/2012/08/16/with-3-million-users-getglue-goes-big-with-a-new-social-tv-app-built-just-for-the-ipad/.

63. Chris Welch, "Social Network GetGlue Revamps iPhone App with 'Hyper-Personalized' Content Guide," *The Verge*, September 5, 2013, http://www.theverge.com/web/2013/9/5/4697322/getglue-revamps-iphone-app-with-new-guide-tv-movie-recommendations.

64. Dubois, "GetGlue."

65. "Full Sticker FAQ."

66. Welch, "GetGlue Revamps."

67. Ken Yeung, "GetGlue Partners with Gnip to Sell Its Second-Screen Entertainment Data," *TheNextWeb*, July 31, 2013, http://thenextweb.com/insider/2013/07/31/getglue-partners-with-gnip-to-sell-its-second-screen-entertainment-data/.

68. Edelsburg, "GetGlue Announces Change to Sticker Program, Fans React," *Lost Remote*, December 4, 2013, http://www.adweek.com/lostremote/getglue-announces-change-to-sticker-program-fans-react/40779.

69. Ryan Lawler, "Miso Updates Social TV App, Nabs *Bridezillas* Partnership," *Gigaom*, August 12, 2010, https://gigaom.com/2010/08/12/miso-updates-social-tv-app-nabs-bridezillas-partnership/.

70. "Want More Than Just Badges? Introducing . . . Fan Club!" *Go Miso*, June 25, 2010, https://gomiso.wordpress.com/2010/06/25/want-more-than-just-badges-introducing-fan-club/#more-94.

71. Anthony Ha, "Miso Takes Social TV Beyond Badges," *Venture Beat*, June 25, 2010, https://venturebeat.com/2010/06/25/miso-fan-clubs/.

72. Roettgers, "The Future of Social TV: It's Not About the Check-In," *Gigaom*, March 15, 2011, https://gigaom.com/2011/03/15/miso-moves-past-check-ins/.

73. Ha, "Miso Takes Social TV Beyond Check-Ins," *Adweek*, December 14, 2011, http://www.adweek.com/digital/miso-takes-social-tv-beyond-check-ins-137098/.

74. Ha, "Miso Takes Social TV Beyond Check-Ins."

75. Anna Heim, "How Miso's CEO Plans to Make TV A More Social Experience," *TheNextWeb*, December 22, 2011, https://thenextweb.com/media/2011/12/22/how-misos-ceo-plans-to-make-tv-a-more-social-experience/.

76. José van Dijck and David Nieborg, "Wikinomics and Its Discontents: A Critical Analysis of Web 2.0 Business Manifestos," *New Media & Society* 11, no. 5 (2009): 856.

77. Heim, "How Miso's CEO Plans."

78. Will Straw, "The Circulatory Turn," in *The Wireless Spectrum: The Politics, Practices, and Poetics of Mobile Media*, eds. Barbara Crow, Michael Longford, and Kim Sawchuk (Toronto: University of Toronto Press, 2010), 28.

79. Chad Elkins, SideShow for *Falling Skies*, "A More Perfect Union," *Miso*, accessed December 3, 2012, http://show.gomiso.com/w8lXHu2#2dGws8G, dead link.

80. Telephone interview with Andrew Seroff, March 15, 2017.

81. Interview with Seroff.

82. Mirko Schäfer, *Bastard Culture! How User Participation Transforms Cultural Production* (Amsterdam: Amsterdam University Press, 2011), 44.

83. Proulx and Shepatin, *Social TV*, 66.

84. Lacy, "Quips."

85. Interview with Seroff.

86. Interview with Seroff.

87. Lawler, "Social TV Miso Pivots Again, Launches Quips App to Let Users Share Their Favorite TV Moments," *TechCrunch*, November 5, 2012, https://techcrunch.com/2012/11/05/miso-quips/.

88. Prakash Venkataraman, "Miso Launches Quips, The Best Way to Talk About TV," *Go Miso*, October 30, 2012, http://blog.gomiso.com:80/2012/10/30/miso-launches-quips-the-best-way-to-talk-about-tv/.

89. Miso, "Introducing Quips for iPhone," Vimeo video, October 30, 2012, https://vimeo.com/52719782.

90. Interview with Seroff.

91. Aseroff, on *Homeland*, "Two Hats," *Miso Quips*, accessed November 26, 2012, no permalink.

92. RyanLawler, on *Breaking Bad*, "Say My Name," *Miso Quips*, accessed November 26, 2012, no permalink.

93. Shawnleiker, on *Fringe*, "Five-Twenty-Ten," *Miso Quips*, accessed November 26, 2012, no permalink.

94. Brokenfat3, on *The Walking Dead*, "When the Dead Coming Knocking," *Miso Quips*, accessed November 26, 2012, no permalink.

95. Graeme Turner, "'Liveness' and 'Sharedness' Outside the Box," *Flow*, April 8, 2011, http://www.flowjournal.org/2011/04/liveness-and-sharedness-outside-the-box/.

96. Lawler, "Social TV Miso Pivots Again."

97. Interview with Seroff.

98. Interview with Seroff.

99. Roettgers, "Dijit Buys Miso's Assets as Social TV Space Consolidates," *Gigaom*, February 1, 2013, https://gigaom.com/2013/02/01/dijit-acquires-miso/.

100. Interview with Seroff.

101. Interview with Seroff.

102. Dumenco, "Loyalty Program."

103. Thomas, "How Viggle Pays People for Watching TV," *Slate*, May 25, 2012, http://www.slate.com/blogs/browbeat/2012/05/25/get_paid_to_watch_tv_viggle_offers_gift_cards_and_more_to_tv_watchers_.html.

104. Thomas, "How Viggle Pays People for Watching TV."

105. van Dijck, "Users Like You," 49.

106. Dumenco, "Loyalty Program."

107. Joshua R. Weaver, "The New Age of Second Screen: Enabling Interaction," *Ad Monsters*, June 5, 2013, http://www.admonsters.com/blog/second-screen-enabling-interaction.

108. Rimma Kats, "DirecTV Teams up with Viggle to Reward Television Viewers," *Mobile Marketer*, August 31, 2012, http://www.mobilemarketer.com/ex/mobilemarketer/cms/news/television/13670.html.

109. Dumenco, "Loyalty Program."

110. Jeanine Poggi, "Social TV Moves Beyond Promotional Role, Becomes Content in Own Right," *Ad Age*, May 9, 2012, http://adage.com/article/special-report-social-tv-conference/social-tv-moves-promotional-role-content/234648/; Adam Flomenbaum, "Viggle Gives Advertisers 'Access' to Highly Targeted Second Screen Audience," *Lost Remote*, August 18, 2014, http://www.adweek.com/lostremote/viggle-gives-advertisers-access-to-highly-targeted-second-screen-audience/46863.

111. Interview with Seroff.

112. Dumenco, "Loyalty Program."

113. Daniel Chamberlin, "Media Interfaces, Networked Media Spaces, and the Mass Customization of Everyday Space," in *Flow TV: Television in the Age of Modern Convergence*, eds. Michael Kackman et al. (New York: Routledge, 2011), 20–21.

114. Poggi, "Social TV Moves Beyond."

115. Poggi, "Watch: Social TV App Viggle to Air Its First TV Commercial," *Ad Age*, March 20, 2013, http://adage.com/article/media/watch-social-tv-app-viggle-air-tv-commercial/240433/.

116. Viggle, "I Watch TV for a Living," YouTube video, 2013, http://www.youtube.com/watch?v=gqejSHgX6ik&feature=youtube_gdata_player, dead link.

117. Thomas, "How Viggle Pays"; Mike Tuttle, "Get Paid to Watch TV," *Web Pro News*, January 26, 2012, http://www.webpronews.com/get-paid-to-watch-tv-2012-01/; Rebecca Grant, "Stop Working and Watch TV: Viggle Hits 1M Users, Opens Up Code to Developers," *Venture Beat*, July 31, 2012, https://venturebeat.com/2012/07/31/stop-working-and-watch-tv-viggle-hits-1m-users-opens-up-code-to-developers/.

118. Christian Fuchs, "Labor in Informational Capitalism and On the Internet," *The Information Society* 26, no. 3 (2010): 191.

119. Sam Thielman, "Viggle Acquires GetGlue," *Adweek*, November 19, 2012, http://www.adweek.com/tv-video/viggle-acquires-getglue-145293/.

120. Baumgartner, "Ad Bridge."

121. Interview with Seroff.

122. Interview with Seroff.

123. Anne Freier, "Perk.com Acquires Viggle to Expand Reward and Ad Options," *Moby Affiliates*, December 15, 2015, http://www.mobyaffiliates.com/blog/perk-com-acquires-viggle-to-expand-mobile-rewards-and-ad-options/.

124. Dumenco, "Social TV is Dead; Long Live Social TV," *Ad Age*, May 12, 2014, http://adage.com/article/the-media-guy/social-tv-dead-long-live-social-tv/293115/.

125. Interview with Seroff.

126. Beth, "GetGlue Issues with Fan/Superfan Stickers," *GetGlue Stickers*, August 9, 2011, http://getgluestickers.com/showthread.php?t=1008; Tech Paradox, "Sticker Hunting Tips," *GetGlue Sticker FAQ*, February 3, 2011, http://stickerfaq.blogspot.com/2011/02/sticker-hunting-tips.html; "Viggle Tips," *Vigglers*, http://vigglers.com/about/.

127. Interview with Seroff.

128. Pierre Lévy, *Collective Intelligence: Mankind's Emerging World in Cyberspace*, trans. Robert Bononno (New York: Plenum, 1997); Henry Jenkins, *Fans, Bloggers, and Gamers: Exploring Participatory Culture* (New York: NYU Press, 2006).

CHAPTER 4

1. Michael Schneider, "Q&A: Amazon Studios Chief on How His Pilot Face-Off Will Work," *TV Guide*, April 19, 2013, http://www.tvguide.com/news/amazon-studios-chief-1064266.aspx; David Lieberman, "Amazon Studios Will Develop Comedy and Children's Series," *Deadline*, May 2, 2012, http://www.deadline.com/2012/05/amazon-studios-will-develop-comedy-and-childrens-series.

2. Drew Grant, "Amazon Outsourcing Its Pilot Season to Viewers," *New York Observer*, February 7, 2014, http://observer.com/2014/02/amazon-outsourcing-its-pilot-season-to-viewers-video/; Nikki Finke, "That Amazon Studios Screenplay Contest: Heavenly or Hellish? Scribes Weigh In," *Deadline*, November 24, 2010, http://www.deadline.com/2010/11/that-amazon-studios-screenplay-contest-heavenly-or-hellish-scribes-weigh-in/; "Amazon Adding Scripts to Its Cart," *Variety*, November 17, 2010, 1; Amazon Studios, "Pilot Season," April 19, 2013, accessed via the Web Archive, https://web.archive.org/web/20130420153759/http://www.amazon.com/gp/feature.html?ie=UTF8&docId=100115558.

3. Amazon Studios, "Getting Started & FAQ," November 2010, accessed via the Web Archive, https://web.archive.org/web/20101120214514/http://studios.amazon.com/getting-started.

4. Peter Kafka, "Amazon Shows Off Its First TV Shows, and Wants to Know What You Think," *All Things Digital*, April 19, 2013, http://allthingsd.com/20130419/amazon-shows-off-its-first-tv-shows-and-wants-you-to-know-what-you-think/.

5. Matt Hills, "*Veronica Mars*, Fandom, and the 'Affective Economics' of Crowdfunding," *New Media & Society* 17, no. 2 (2015): 183–97.

6. Daren C. Brabham, "Crowdsourcing as a Model for Problem Solving: An Introduction and Cases," *Convergence* 14, no. 1 (2008): 75–90.

7. Axel Bruns, *Blogs, Wikipedia, Second Life, and Beyond: From Production to Produsage* (New York: Peter Lang), 2008; José van Dijck, "Users Like You? Theorizing Agency in User-Generated Content," *Media, Culture & Society* 31, no. 1 (2009): 41–58; John Banks and Sal Humphreys, "The Labor of User Co-Creators: Emergent Social Network Markets?" *Convergence* 14.4 (2008): 401–18.

8. Brabham, "The Myth of Amateur Crowds: A Critical Discourse Analysis of Crowdsourcing Coverage," *Information, Communication & Society* 15, no. 3 (2012): 405–6.

9. Simone Murray, "'Celebrating the Story the Way It Is': Cultural Studies, Corporate Media, and the Contested Utility of Fandom," *Continuum* 18, no. 1 (2004): 7–25.

10. Henry Jenkins, Sam Ford, and Joshua Green, *Spreadable Media: Creating Meaning and Value in a Networked Culture*. (New York: NYU Press, 2013).

11. Will Brooker, "Going Pro: Gendered Responses to the Incorporation of Fan Labor as User-Generated Content," in *Wired TV: Laboring Over an Interactive Future*, eds. Denise Mann (New Brunswick: Rutgers University Press, 2014), 72–97.

12. Tiziana Terranova, "Free Labor: Producing Culture for the Digital Economy," *Social Text* 18, no. 2 (2004): 33–58.

13. Brooker, "Going Pro," 90–91. Mark Andrejevic, "Exploitation in the Data Mine," in *Internet and Surveillance: The Challenges of Web 2.0 and Social Media*, eds. Christian Fuchs et al. (New York: Routledge, 2012), 79.

14. Maurizio Lazzarato, "Immaterial Labor," in *Radical Thought in Italy: A Potential Politics*, eds. Paulo Virno and Michael Hardt (Minneapolis: University of Minnesota Press, 1996), 133–50.

15. James Beniger, *The Control Revolution* (Cambridge: Harvard University Press, 1986).

16. Nathan McAlone, "Amazon CEO Jeff Bezos Said Something about Prime Video That Should Scare Netflix," *Business Insider*, June 2, 2016, http://www.businessinsider.com/amazon-ceo-jeff-bezos-said-something-about-prime-video-that-should-scare-netflix-2016-6.

17. "'Amazon Unbox on TiVo' Now Available, Offering 1.5 Million Broadband-Ready TiVo Subscribers Access to Thousands of Movies and TV Shows," March 7, 2007, http://phx.corporate-ir.net/phoenix.zhtml?c=97664&p=irol-newsArticle&ID=971365.

18. "Amazon Customers Can Now Instantly Watch Ad-Free Movies and TV Shows on Macs, PCs and Compatible Sony BRAVIA Televisions Starting Today on Amazon Video on Demand," September 3, 2008, http://phx.corporate-ir.net/phoenix.zhtml?c=176060&p=NewsArticle&id=1193455.

19. "Amazon Prime Members Now Get Unlimited, Commercial-free, Instant Streaming of More Than 5,000 Movies and TV Shows at No Additional Cost," February 22, 2011, http://phx.corporate-ir.net/phoenix.zhtml?c=176060&p=irol-newsArticle&ID=1531234.

20. Nick Wingfield and Brian Stelter, "How Netflix Lost 800,000 Members, and Goodwill," *New York Times*, October 24, 2011, http://www.nytimes.com/2011/10/25/technology/netflix-lost-800000-members-with-price-rise-and-split-plan.html.

21. "Amazon Adds Movies to Streaming Service to Challenge Netflix," *Ad Age*, September 4, 2012, http://adage.com/article/media/amazon-adds-epix-movies-streaming-service/237003/.

22. Amazon Prime Video, "Welcome Screenwriters and Filmmakers," YouTube video, August 18, 2011, https://www.youtube.com/watch?v=pTcBKBKbV5I.

23. "Amazon.com Launches Amazon Studios—A New Online Business to Discover New Talent and Develop Motion Pictures," November 16, 2010, http://phx.corporate-ir.net/phoenix.zhtml?c=176060&p=irol-newsArticle&ID=1497259.

24. Todd Gitlin, *Inside Prime Time* (New York: Pantheon, 1983), 32–45.

25. For instance, see Elizabeth Wagmeister, "Early Pilot Buzz: What's Heating Up for the 2017–2018 TV Season?" *Variety*, April 28, 2017, http://variety.com/2017/tv/news/pilot-buzz-2017-2018-television-season-1202393937/.

26. Michele Willens, "Putting Films to the Test, Every Time," *New York Times*, June 25, 2000, http://www.nytimes.com/2000/06/25/movies/film-putting-films-to-the-test-every-time.html.

27. Ryan Koo, "Is Amazon Studios the Future of Film or Is It a Bastardization of Crowdsourcing?" *No Film School*, November 17, 2010, https://nofilmschool.com/2010/11/amazon-studios-future-film-travesty; Mark Brown, "Amazon Studios Democratizes Moviemaking," *Wired*, November 18, 2010, https://www.wired.com/underwire/2010/11/amazon-studios/.

28. Finke, "Amazon Studios Screenplay Contest."

29. Ellen E. Jones, "The Amazon Movie Revolution . . . One Year Later," *The Guardian*, December 1, 2011, https://www.theguardian.com/film/2011/dec/01/amazon-studios-revolution.

30. Joan E. Solsman, "How Amazon Studios Went from Grassroots Idealist to Hollywood Threat," *CNET*, November 14, 2013, https://www.cnet.com/news/how-amazon-studios-went-from-grassroots-idealist-to-hollywood-threat/.

31. Jones, "Amazon Movie Revolution."

32. "Amazon to Stream Vids," *Variety*, February 2, 2011, 5.

33. Greg Sandoval, "Netflix CEO on Amazon Prime Video: 'A Confusing Mess,'" *CNET*, September 26, 2012, https://www.cnet.com/news/netflix-ceo-on-amazon-prime-video-a-confusing-mess/.

34. Cynthia Littleton, "Netflix Originals Vision Catches Biz's Eye," *Variety*, March 20, 2012, 2.

35. Brian Stelter, "A Drama's Streaming Premiere," *New York Times*, February 18, 2013, http://www.nytimes.com/2013/01/20/arts/television/house-of-cards-arrives-as-a-netflix-series.html.

36. Andrew Wallenstein, "Amazon Takes on TV," *Variety*, May 3, 2012, 4.

37. Solsman, "Grassroots Idealist."

38. Schneider, "Q&A: Amazon Studios Chief."

39. Solsman, "Grassroots Idealist."

40. Ryan Tate, "Why He Ditched Posh Hollywood Perks: 10 Questions with Amazon Studios Chief Roy Price," *Wired*, November 11, 2012, http://www.wired.com/2012/11/amazon-studios-roy-price.

41. "Prime Instant Video Greenlights *Betas*, an Original Comedy Pilot," March 27, 2013, http://phx.corporate-ir.net/phoenix.zhtml?c=176060&p=irol-newsArticle&ID=1801168.

42. Natalie Jarvey, "Amazon Studios Head Roy Price on Competing with Netflix, Xbox Studios's Demise (Q&A)," *Hollywood Reporter*, July 30, 2014, http://www.hollywoodreporter.com/news/amazon-studios-head-roy-price-721867.

43. Andrejevic, *iSpy: Surveillance and Power in the Interactive Era* (Lawrence: University Press of Kansas, 2007), 27.

44. Schneider, "Q&A: Amazon Studios Chief."

45. Schneider, "Q&A: Amazon Studios Chief."

46. Derek Johnson, "Inviting Audiences In: The Spatial Reorganization of Production and Consumption in 'TVIII,'" *New Review of Film and Television Studies* 5, no. 1 (2007): 63.

47. Jarvey, "Amazon Studios Announces Second Slate of Pilots," *Hollywood Reporter*, February 6, 2014, http://www.hollywoodreporter.com/news/amazon-studios-announces-second-slate-677646.

48. Nico Carpentier, *Media and Participation: A Site of Ideological-Democratic Struggle* (Chicago: Intellect, 2011), 215.

49. Banks and Humphreys, "Labor of User Co-Creators," 412.

50. "Prime Instant Video Greenlights."

51. "Amazon Studios Debuts New Original Pilots Today," February 6, 2014, http://phx.corporate-ir.net/phoenix.zhtml?c=176060&p=irol-newsArticle&ID=1897711.

52. Bradford Evans, "Ranking Amazon's 8 New Comedy Pilots, from Worst to First," *Splitsider*, April 22, 2013, http://splitsider.com/2013/04/ranking-amazons-8-new-comedy-pilots-from-worst-to-first/; Carl Franzen, "Amazon Debuts All 14 Original Series Pilots for Free Streaming," *The Verge*, April 19, 2013, https://www.theverge.com/2013/4/19/4236092/amazon-studios-debuting-14-original-pilots-free-streaming.

53. Mel Stanfill, *Exploiting Fandom: How the Media Industry Seeks to Manipulate Fans* (Iowa City: University of Iowa Press, 2019), 19.

54. Ted Striphas, "Algorithmic Culture," *European Journal of Cultural Studies* 18, nos. 4–5 (2015): 407.

55. Schneider, "Q&A: Amazon Studios Chief."

56. van Dijck, "Users Like You?" 49.

57. "Six More Amazon Original Pilots Receive Greenlight for Full Seasons," March 31, 2014, http://phx.corporate-ir.net/phoenix.zhtml?c=176060&p=irol-newsArticle&ID=1913643.

58. Spangler, "Amazon Greenlights 6 Series, Including Chris Carter's 'The After,' and Renews 'Alpha House,'" *Variety*, March 31, 2014, http://variety.com/2014/digital/news/amazon-greenlights-6-series-including-chris-carters-the-after-and-renews-alpha-house-1201150016/; Myles McNutt, "This Is the Review Data for the Latest Round of Amazon Pilots around the Time They Went Offline Last Night," Tumblr post, March 11, 2014, http://mylesmcnutt.tumblr.com/post/79173485238/via-neogaf-user-returnoftherat-this-is-the-review.

59. "Amazon Studios Debuts New Original Pilots Today," February 6, 2014, http://phx.corporate-ir.net/phoenix.zhtml?c=176060&p=irol-newsArticle&ID=1897711.

60. June Thomas, "Can Amazon Transform TV?" *Slate*, April 22, 2013, http://www.slate.com/articles/arts/television/2013/04/amazon_tv_pilots_john_goodman_and_bebe_neuwirth_star_in_new_streaming_sitcoms.html; Hillary Atkin, "Disrupters: The Execs Are Rocking the Business of Content Production, Distribution, and Consumption (Roy Price)," *Variety*, December 10, 2012, http://variety.com/2012/digital/markets-festivals/roy-price-director-amazon-studios-1118063170/.

61. Robert J. Thompson, "Preface," in *Quality TV: Contemporary American Television and Beyond*, eds. Janet McCabe and Kim Akass (New York: I. B. Tauris, 2007), xvii–xx.

62. Jane Feuer, "HBO and the Concept of Quality TV," in *Quality TV: Contemporary American Television and Beyond*, eds. Janet McCabe and Kim Akass (New York: I. B. Tauris, 2007), 147.

63. Janet McCabe and Kim Akass, "Introduction: Debating Quality," in *Quality TV: Contemporary American Television and Beyond*, eds. Janet McCabe and Kim Akass (New York: I. B. Tauris, 2007), 3.

64. Gary Edgerton and Jeffrey P. Jones, eds., *The Essential HBO Reader* (Lexington: University Press of Kentucky, 2009); Amanda D. Lotz, *Cable Guys: Television and Masculinities in the 21st Century* (New York: NYU Press, 2014); Taylor Nygaard and Jorie Lagerwey, "Broadcasting Quality: Re-Centering Feminist Discourse with *The Good Wife*," *Television & New Media* 18, no. 2 (2017): 105–13.

65. Deborah L. Jaramillo, "The Family Racket: AOL Time Warner, HBO, *The Sopranos*, and the Construction of a Quality Brand," *Journal of Communication Inquiry* 26, no. 1 (2002): 59–75; Catherine Johnson, "Tele-Branding in TVIII: The Network as Brand and the Programme as Brand," *New Review of Film and Television Studies* 5, no. 1 (2007): 5–24.

66. Dean J. DeFino, *The HBO Effect* (New York: Bloomsbury, 2013); Taylor Nygaard, "Girls Just Want to be 'Quality': Lena Dunham, and *Girls*' Conflicting Brand Identity," *Feminist Media Studies* 13, no. 2 (2013): 253–62.

67. Robin Nelson, "'Quality Television': '*The Sopranos* Is the Best Television Drama Ever . . . in My Humble Opinion,'" *Critical Studies in Television* 1, no. 1 (2006): 58–71.

68. Michael Z. Newman and Elana Levine, *Legitimating Television: Media Convergence and Cultural Status* (New York: Routledge, 2011), 7.

69. Christopher Anderson, "Producing an Aristocracy of Culture in American Television," in *The Essential HBO Reader*, eds. Gary R. Edgerton and Jeffrey P. Jones (Lexington: University Press of Kentucky, 2009), 38.

70. Joe Adalian, "Heading into Its Sophomore TV Season, Amazon Takes a Big Step Forward," *Vulture*, March 12, 2014, http://www.vulture.com/2014/03/amazon-gets-ready-for-a-big-sophomore-season.html; Erik Adams et al., "Amazon's Latest Pilots Are a Cut above Its Last Batch—and Most Other Pilots," *The A. V. Club*, February 12, 2014, http://www.avclub.com/review/amazons-latest-pilots-are-a-cut-above-its-last-bat-201124.

71. Willa Paskin, "Amazon Has Finally Made Its *House of Cards*," *Slate*, February 11, 2014, http://www.slate.com/blogs/browbeat/2014/02/11/amazon_s_new_pilots_transparent_mozart_in_the_jungle_the_after_the_rebels.html.

72. "Six More Amazon Original Pilots Receive Greenlight."

73. Jarvey, "Amazon Studios's Roy Price Reveals Series Orders."

74. Newman and Levine, *Legitimating Television*, 39.

75. Michele Hilmes, *Only Connect: A Cultural History of Broadcasting in the United States* (Belmont: Wadsworth, 2002), 305.

76. Nygaard, "Girls Just Want to be 'Quality.'"

77. Deadline Team, "TCA: Amazon's 'Transparent' Cast Bristle at Talk of Taking Less Pay for Creative Freedom," *Deadline*, July 12, 2014, http://deadline.com/2014/07/tca-amazons-transparent-cast-bristle-at-talk-of-taking-less-pay-for-creative-freedom-803132.

78. Spangler, "Amazon's Test after 'Transparent': Can Studio Maintain Its Momentum?" *Variety*, September 2, 2014, http://variety.com/2014/digital/news/amazons-test-after-transparent-can-studio-maintain-its-momentum-1201307545/.

79. Philiana Ng, "'Transparent' Team Talks Binge Viewing, Defends Digital Platform Play," *Hollywood Reporter*, July 12, 2014, http://www.hollywoodreporter.com/live-feed/transparent-team-talks-binge-viewing-718157.

80. Shelli Weinstein, "It's a Collaborative Effort: Amazon Stars Give Five Stars on the Comments Section," *Variety*, September 16, 2014, http://variety.com/2014/digital/news/amazon-studios-pilot-stars-comments-1201306774/.

81. Feuer, "HBO and the Concept of Quality TV"; Robin Nelson, "'Quality Television'"; Nygaard, "Girls Just Want to Be 'Quality.'"

82. Thompson, "Preface," xiii.

83. Jarvey, "Roy Price on Competing with Netflix."

84. Weinstein, "Amazon's 'Transparent' Cast Talks Family, Making a Difference at Series Premiere," *Variety*, September 16, 2014, http://variety.com/2014/scene/news/amazons-transparent-cast-talks-family-making-a-difference-at-series-premiere-1201306758/.

85. Sarah Caldwell, "Is Quality TV Any Good? Generic Distinctions, Evaluations and the Troubling Matter of Critical Judgment," in *Quality TV: Contemporary American Television and Beyond*, eds. Janet McCabe and Kim Akass (New York: I. B. Tauris, 2007), 19–34.

86. Jason Lynch, "How Amazon Built a TV Studio That's Finally Challenging Netflix," *QZ*, August 12, 2014, http://qz.com/246925/how-amazon-built-a-tv-studio-thats-finally-challenging-netflix/.

87. Spangler, "Amazon's Test after *Transparent*."

88. "Watch on Amazon Instant Video Today: Amazon Debuts Five New Original Comedy and Drama Pilots," August 28, 2014, http://phx.corporate-ir.net/phoenix.zhtml?c=176060&p=irol-newsArticle&ID=1961787.

89. McNutt, "This Is the Review Data."

90. RatskyWatsky, post on NeoGAF message board, September 21, 2014, https://www.neogaf.com/threads/amazon-releases-5-new-tv-pilots.878435/post-130964705.

91. "Amazon Originals *Hand of God* and *Red Oaks* Receive Greenlight for Full Seasons," October 3, 2014, http://phx.corporate-ir.net/phoenix.zhtml?c=176060&p=irol-newsArticle&ID=1973838.

92. Nellie Andreeva, "Chris Carter Drama Series 'The After' Not Going Forward at Amazon Studios," *Deadline*, January 5, 2015, http://deadline.com/2015/01/the-after-chris-carter-series-cancelled-amazon-studios-1201341268/.

93. Adalian, "What the HBO-Amazon Deal Means for Netflix and 5 Other Questions Answered," *Vulture*, April 24, 2014, http://www.vulture.com/2014/04/hbo-amazon-deal-netflix-questions.html.

94. Scott Porch, "A Year After Launch, Amazon Channels Has Grown from Start-Up Service to Full-Blown Platform," *Decider*, December 16, 2016, https://decider.com/2016/12/16/a-year-after-launch-amazon-channels-has-grown-from-a-service-to-a-full-blown-platform/.

95. Alisa Perren, *Indie, Inc.: Miramax and the Transformation of Hollywood in the Late 1990s* (Austin: University of Texas Press, 2012), 17.

96. Perren, *Indie, Inc.*, 16–53.

97. Mike Fleming Jr., "Deadline Eyeballs Weinstein TV Sale Prospectus: So That's Where the Money Is Going," *Deadline*, January 12, 2017, http://deadline.com/2017/01/deadline-eyeballs-weinstein-tv-sale-prospectus-so-thats-where-the-money-is-going-1201883459/.

98. Brian Formo, "How Amazon Is Operating Like a 1970s Studio," *Collider*, December 21, 2016, http://collider.com/amazon-studios-oscars-manchester-by-the-sea/.

99. Brent Lang, "Amazon, Retail Behemoth, Taking Small Steps into Hollywood," *Variety*, March 8, 2016, http://variety.com/2016/digital/features/amazon-studios-movies-marketing-sxsw-1201724314/.

100. McNutt, "Amazon Pilot Season—January 2015—By the Numbers," Tumblr post, February 18, 2015, http://mylesmcnutt.tumblr.com/post/111384004686/amazon-pilot-season-january-2015-by-the; RatskyWatsky, post on the NeoGAF message board, November 18, 2015, https://www.neogaf.com/threads/amazon-pilot-season-watch-and-vote-now.1135595/post-185974175.

101. John Lynch, "Amazon is Reportedly on the Hook for 4 More Woody Allen Movies, and Considering a 'Hefty Payout' to Kill the Deal," *Business Insider*, January 31, 2018, http://www.businessinsider.com/amazon-considering-hefty-payout-to-end-movie-deal-with-woody-allen-report-2018-1.

102. Adam Sternbergh, "How Amazon Became the Moneyball Network," *Vulture*, May 6, 2015, http://www.vulture.com/2015/05/amazon-tv-moneyball-network.html.

103. Caroline Framke, "Amazon's 'Democratic' TV Development Process May Have a Sexist Loophole," *Vox*, March 7, 2017, https://www.vox.com/culture/2017/3/7/14795108/amazon-original-series-orders-amy-sherman-palladino.

104. Lesley Goldberg, "'Good Girls Revolt' Canceled at Amazon; Will Be Shopped Elsewhere," *Hollywood Reporter*, December 2, 2016, https://www.hollywoodreporter.com/live-feed/good-girls-revolt-canceled-at-amazon-will-be-shopped-952149. Dino Day-Ramos, "'The Last Tycoon' Canceled at Amazon after One Season," *Deadline*, September 9, 2017, http://deadline.com/2017/09/the-last-tycoon-matt-bomer-billy-ray-f-scott-fitzgerald-amazon-1202165509/.

105. Spangler, "Amazon's Big NFL Play Could Kick Off a Shake-Up in TV Sports Rights," *Variety*, September 26, 2017, http://variety.com/2017/digital/news/amazon-nfl-thursday-night-football-sports-rights-1202570293/; Andreeva, "Amazon Sets 'The Lord of the Rings' Trilogy Series in Mega Deal with Multi-Season Commitment," *Deadline*, November 13, 2017, http://deadline.com/2017/11/amazon-the-lord-of-the-rings-tv-series-multi-season-commitment-1202207065/.

106. Sternbergh, "How Amazon Became the Moneyball Network."

107. Cynthia Littleton and Daniel Holloway, "Jeff Bezos Mandates Programming Shift at Amazon Studios," *Variety*, September 8, 2017, http://variety.com/2017/tv/news/amazon-studios-jeff-bezos-roy-price-zelda-1202552532/.

108. Littleton and Holloway, "Jeff Bezos Mandates Programming Shift."

109. Kim Masters and Lesley Goldberg, "Amazon Studios's Roy Price: Inside the Fall of a Top Executive (and What's Next)," *Hollywood Reporter*, October 18, 2017, https://www.hollywoodreporter.com/features/amazon-studios-roy-price-inside-fall-a-top-executive-whats-next-1049859; Ben Fritz and Joe Flint, "Where Amazon Is Failing to Dominate:

Hollywood," *Wall Street Journal*, October 6, 2017, https://www.wsj.com/articles/where-amazon-is-failing-to-dominate-hollywood-1507282205.

110. Fritz and Flint, "Where Amazon Is Failing."

111. Framke, "Amazon's 'Democratic' TV Development Process."

112. Masters and Goldberg, "Inside the Fall; Masters, "Behind Amazon Studios's Executive Purge as New Details Emerge," *Hollywood Reporter*, November 1, 2017, https://www.hollywoodreporter.com/features/behind-amazon-studios-executive-purge-as-new-details-emerge-1053579.

113. "Amazon's Fall 2017 Half-Hour Pilot Season Premieres on November 10 on Amazon Prime Video," press release, October 23, 2017, http://phx.corporate-ir.net/phoenix.zhtml?c=176060&p=irol-newsArticle&ID=2310369.

114. Goldberg, "Amazon Passes on Trio of Comedy Pilots," *Hollywood Reporter*, December 18, 2017, https://www.hollywoodreporter.com/live-feed/amazon-passes-trio-comedy-pilots-1068862.

115. Andrew R. Chow, "Amazon Cancels 'One Mississippi,' 'I Love Dick,' and 'Jean-Claude,'" *New York Times*, January 18, 2018, https://www.nytimes.com/2018/01/18/arts/television/amazon-cancels-one-mississippi-i-love-dick-and-jean-claude.html.

116. Meg James, "Amazon Studios Cuts Ties with Weinstein Co. Following Harvey Weinstein Sex Scandal," *Los Angeles Times*, October 13, 2017, http://www.latimes.com/business/hollywood/la-fi-ct-amazon-cuts-ties-weinstein-co-20171013-story.html; Lynch, "Amazon is Reportedly on the Hook."

CHAPTER 5

1. For more on Twitter's challenges during this time, see Charlie Warzel, "A Honeypot for Assholes": Inside Twitter's 10-Year Failure To Stop Harassment," *BuzzFeed*, August 11, 2016, https://www.buzzfeed.com/charliewarzel/a-honeypot-for-assholes-inside-twitters-10-year-failure-to-s?utm_term=.viedx1lr5#.tgN7N2dV0; Nathan Olivarez-Giles, "Why Twitter Can't Shake Its Harassment Problem," *Wall Street Journal*, August 24, 2016, https://www.wsj.com/articles/why-twitter-cant-shake-its-harassment-problem-1472045224.

2. For more on the range of corporate tweeting, see Alice Truong, "This Twitter Account Shames Brands for Lazy, Awkward, and Awful Tweets," *Fast Company*, October 29, 2013, https://www.fastcompany.com/3020851/this-twitter-account-shames-brands-for-lazy-awkward-and-awful-tweets; Erica Futterman, "DiGiorno Pizza Live-Tweeted NBC's 'The Sound of Music' in 'Unplanned Event' and It Was Amazing," *BuzzFeed*, December 5, 2013, https://www.buzzfeed.com/ericafutterman/digiorno-pizza-live-tweeted-nbcs-the-sound-of-music-in-unpla?utm_term=.um87GLBPx#.hgnwMB8YW; and Taylor Wofford, "The Good, The Bad, and The Ugly of 9/11 Brand Tweets," *Newsweek*, September 11, 2014, http://www.newsweek.com/good-bad-and-ugly-911-brand-tweets-269992.

3. Sarah Banet-Weiser, *Authentic™: The Politics of Ambivalence in a Brand Culture* (New York: NYU Press 2013), 37.

4. Ethan Tussey, *The Procrastination Economy: The Big Business of Downtime* (New York: NYU Press, 2018), 37.

5. Nancy Hass, "And the Award for the Next HBO Goes To . . . ," *GQ*, January 29, 2013, http://www.gq.com/story/netflix-founder-reed-hastings-house-of-cards-arrested-development.

6. Emily Steel, "HBO Plans New Streaming Service, With Eye on Cord Cutters," *New York Times*, October 15, 2014, http://www.nytimes.com/2014/10/16/business/media/time-warner-chief-to-brief-investors-on-plans-for-growth.html.

7. David Lieberman, "HBO to Launch Stand-Alone Online Service, Without Cable, In 2015: Time Warner Investor Day," *Deadline*, October 15, 2014, http://deadline.com/2014/10/richard-plepler-time-warner-investor-day-hbo-ceo-presentation-851815/.

8. For other examples of how scholars classify and organize tweets see Arkaitz Zubiaga et al., "Real-Time Classification of Twitter Trends," *Journal of the Association of Information Science & Technology* 66, no. 3 (2015): 462–73; and Andreas Kanavos et al., "Large Scale Implementations for Twitter Sentiment Classification," *Algorithms* 10, no. 1 (2017): 1–21.

9. Catherine Johnson, *Branding Television* (New York: Routledge, 2012), 20–21.

10. Naomi Klein, *No Logo: Taking Aim at the Brand Bullies* (New York: Picador, 2002), 22.

11. Klein, *No Logo*, 22.

12. Charles R. Acland, *Screen Traffic: Movies, Multiplexes, and Global Culture* (Durham: Duke University Press, 2003), 23–25.

13. Martine Danan, "Marketing the Hollywood Blockbuster in France," *Journal of Popular Film and Television* 23, no. 2 (1995): 131–40; Justin Wyatt, "The Formation of the Major Independent: Miramax, New Line, and the New Hollywood," in *Contemporary Hollywood Cinema*, eds. Steve Neale and Murray Smith (London: Routledge, 1998), 74–90; John McMurria, "Long-Format TV: Globalization and Network Branding in a Multi-Channel Era," in *Quality Popular Television: Cult TV, The Industry, and Fans*, eds. Mark Jancovich and James Lyons (London: British Film Institute, 2003), 65–87.

14. Acland, *Screen Traffic*, 79.

15. For more on cable and cable branding, see Megan Mullen, *The Rise of Cable Programming in the United States: Revolution or Evolution?* (Austin: University of Texas Press, 2009); Kevin Sandler, "'A Kid's Gotta Do What a Kid's Gotta Do': Branding the Nickelodeon Experience," in *Nickelodeon Nation: The History, Politics, and Economics of America's Only TV Channel for Kids*, ed. Heather Hendershot (New York: NYU Press, 2004), 62–99; Amanda D. Lotz, *Redesigning Women: Television After the Network Era* (Urbana: University of Illinois Press, 2006).

16. Johnson, *Branding Television*, 18. "Brand filter" is a phrase often used within the industry. See Johnnie Roberts, "The Woman Behind USA Network's Hit Machine," *Newsweek*, July 10, 2009, http://www.newsweek.com/woman-behind-usa-networks-hit-machine-81947.

17. Sam Ward, "Investigating the Practices of Television Branding: An Afterword on Methodology," *Networking Knowledge* 7, no. 1 (2014): 55.

18. Johnson, *Branding Television*, 49.

19. Celia Lury, *Brands: The Logos of the Global Economy* (New York: Routledge, 2004), 6; emphasis in original.

20. Adam Arvidsson, *Brands: Meaning and Value in Media Culture* (New York Routledge, 2006), 8, 82.

21. Banet-Weiser, *Authentic*, 9.

22. Arvidsson, "The Ethical Economy: Towards a Post-Capitalist Theory of Value," *Capital and Class*, no. 97 (2009): 17.

23. Christopher Anderson, "Producing an Aristocracy of Culture in American Television," in *The Essential HBO Reader*, eds. Gary Edgerton and Jeffrey P. Jones (Lexington: University Press of Kentucky, 2008), 30; emphasis in original.

24. For instance, see Alan Sepinwall, *The Revolution Was Televised: The Cops, Crooks, Slingers, and Slayers Who Changed TV Drama Forever* (New York: Touchstone, 2012).

25. Avi Santo, "Para-Television and Discourses of Distinction," in *It's Not TV: Watching HBO in the Post-Television Era*, eds. Marc Leverette, Brian L. Ott, and Cara Louise Buckley (New York: Routledge, 2008), 20.

26. Dana Polan, "Cable Watching: HBO, *The Sopranos*, and Discourses of Distinction," *Cable Visions: Television Beyond Broadcasting*, eds. Sarah Banet-Weiser, Cynthia Chris, and Anthony Freitas (New York: NYU Press, 2007), 280.

27. Polan, "Cable Watching," 11.

28. Nicholas Carah, "Algorithmic Brands: A Decade of Brand Experiments with Mobile and Social Media," *New Media & Society* 19, no. 3 (2017): 385.

29. Santo, "Para-Television," 32.

30. Anderson, "Producing an Aristocracy of Culture," 37.

31. Anderson, "Producing an Aristocracy of Culture," 38.

32. Polan, "Cable Watching," 280.

33. Polan, "Cable Watching," 273.

34. Paul Grainge, *Brand Hollywood: Selling Entertainment in a Global Media Age* (New York: Routledge, 2008), 10.

35. Will Brooker, "Living on *Dawson's Creek:* Teen Viewers, Cultural Convergence and Television Overflow," *International Journal of Cultural Studies* 4, no. 4 (2001): 456.

36. Johnson, *Branding Television*, 156.

37. Christina Warren, "HBO Gets social with HBO Connect," CNN, May 25, 2011, http://www.cnn.com/2011/TECH/social.media/05/25/hbo.connect.mashable/index.html.

38. Natan Edelsburg, "A Q&A with Sabrina Caluori about HBO Connect," *Lost Remote*, November 12, 2013, http://www.adweek.com/lostremote/a-qa-with-sabrina-caluori-about-hbo-connect-the-networks-digital-marketing-secret-weapon/40272.

39. Anderson, "Producing an Aristocracy of Culture," 33.

40. Lury, *Brands*, 7–8.

41. Lotz, "If It's Not TV, What Is It? The Case of U.S. Subscription Television," in *Cable Visions: Television Beyond Broadcasting*, eds. Sarah Banet-Weiser, Cynthia Chris, and Anthony Freita (New York: NYU Press, 2007), 95.

42. Santo, "Para-Television," 37.

43. Deadline team, "HBO Go Considered for Non-Cable Subscribers: Report," *Deadline*, March 21, 2013, http://deadline.com/2013/03/hbo-go-considered-for-non-cable-subscribers-report-459382/; Lieberman, "Is It Time for HBO Go to be Offered to People Who Don't

Subscribe to Pay TV?" *Deadline*, March 17, 2014, http://deadline.com/2014/03/is-it-time-for-hbo-go-to-be-offered-to-people-who-dont-subscribe-to-pay-tv-700125/.

44. Lieberman, "HBO Without Cable? Not Yet, Says Time Warner Chief," *Deadline*, May 1, 2013, http://deadline.com/2013/05/hbo-go-without-cable-not-yet-says-time-warner-chief-487633/.

45. Kevin P. Sullivan, "HBO Go Passes 3 Million Downloads. Is It Worth It?" *Entertainment Weekly*, June 27, 2011, http://ew.com/article/2011/06/27/hbo-go-worth-it/.

46. Greg Kumparak, "HBO Doesn't Care If You Share Your HBO Password. For Now.," *TechCrunch*, January 20, 2014, https://techcrunch.com/2014/01/20/hbo-doesnt-care-if-you-share-your-hbo-go-account-for-now/; Kimberly Nordyke, "HBO Go Crashes During 'True Detective' Finale," *Hollywood Reporter*, March 9, 2014, http://www.hollywoodreporter.com/live-feed/hbo-go-crashes-true-detective-687087.

47. Dave Itzkoff, "New 'Arrested Development' Season Coming to Netflix on May 26," *New York Times*, April 4, 2013, https://artsbeat.blogs.nytimes.com/2013/04/04/new-arrested-development-season-coming-to-netflix-on-may-26/; Itzkoff, "Netflix Prison Series Gets July Release Date," *New York Times*, April 30, 2013, https://artsbeat.blogs.nytimes.com/2013/04/30/netflix-prison-series-gets-july-release-date/.

48. Hass, "And the Award for The Next HBO Goes To."

49. Sepinwall, "Why Matt Weiner 'Would Lose' If He Wanted to Make A Weekly Netflix Show," *HitFix,* January 23, 2016, http://www.hitfix.com/whats-alan-watching/ted-talk-state-of-the-netflix-union-discussion-with-chief-content-officer-ted-sarandos.

50. Benny Evangelista, "Cord Cutting Accelerated in 2015, on Track to Continue Next Year," *San Francisco Chronicle*, December 31, 2015, http://www.sfchronicle.com/business/article/Cord-cutting-accelerated-in-2015-on-track-to-6730696.php; Aaron Pressman, "More Than One in Five Households Has Dumped the Gable Goliath," *Fortune*, April 5, 2016, http://fortune.com/2016/04/05/household-cable-cord-cutters/.

51. Zachary M. Seward, "Netflix Now Has More Paying Subscribers than HBO in the United States," *Quartz*, October 21, 2013, https://qz.com/137854/netflix-now-has-more-paying-subscribers-than-hbo-in-the-united-states/; John McDuling, "HBO Doesn't Think Netflix is a Competitor. Here Are All the Reasons That's Wrong," *Quartz*, December 11, 2013, https://qz.com/156551/hbo-doesnt-think-netflix-is-a-competitor-here-are-all-the-reasons-thats-wrong/.

52. Lucas Shaw, "Time Warner CEO Jeff Bewkes: Netflix and HBO Are Not Competitors," *The Wrap*, December 10, 2013, http://www.thewrap.com/time-warner-ceo-dont-compare-netflix-hbo/; John McDuling, "HBO Is Now 'Seriously Considering' Whether to Offer HBO Go without Cable," *Quartz*, September 11, 2014, https://qz.com/263950/hbo-is-now-seriously-considering-whether-to-offer-hbo-go-without-cable-tv/.

53. Steel, "HBO Plans New Streaming Service."

54. Jason Eppink, "A Brief History of The GIF (So Far)," *Journal of Visual Culture* 13, no. 3 (2014): 303.

55. Michael Z. Newman, "GIFs: The Attainable Text," *Film Criticism* 40, no. 1 (2016), https://doi.org/10.3998/fc.13761232.0040.123.

56. Eppink, "A Brief History," 303.

57. Eppink, "A Brief History," 301.

58. Johnson, Branding Television, 48.

59. Polan, "Cable Watching," 273.
60. Tussey, *Procrastination Economy*, 59.
61. Johnson, *Branding Television*, 47–49.
62. Lynn Spigel and Max Dawson, "Television and Digital Media," in *American Thought and Culture in the 21st Century*, eds. Catherine Morley and Martin Halliwell (Edinburgh: Edinburgh University Press, 2008), 281.
63. Grainge, *Brand Hollywood*, 27.
64. Santo, "Para-Television," 27.
65. Carah, "Algorithmic Brands," 385.
66. Carah, "Algorithmic Brands," 385.
67. Carah, "Algorithmic Brands," 398.
68. Brian L. Ott, "Introduction: The Not TV Text," in *It's Not TV: Watching HBO in the Post-Television Era*, eds. Marc Leverette, Brian L. Ott, and Cara Louise Buckley (New York: Routledge, 2008), 99.
69. Ron Becker, *Gay TV and Straight America* (New Brunswick: Rutgers University Press, 2006), 177, 144.
70. For more on corporate rainbow washing, see "The Problem with the Rainbow-Washing of LGBTQ+ Pride," *Wired*, June 21, 2020, https://www.wired.com/story/lgbtq-pride-consumerism/.
71. Roopali Mukherjee and Sarah Banet-Weiser, "Introduction: Commodity Activism in Neoliberal Times," in *Commodity Activism: Cultural Resistance in Neoliberal Times*, eds. Roopali Mukherjee and Sarah Banet-Weiser (New York: NYU Press, 2012), 4, 12.
72. Laurie Ouellette, "Citizen Brand: ABC and the Do Good Turn in US Television," in *Commodity Activism: Cultural Resistance in Neoliberal Times*, eds. Roopali Mukherjee and Sarah Banet-Weiser (New York: NYU Press, 2012), 62.
73. Ouellette, "Citizen Brand," 62, citing Mark Gobé, *Citizen Brand: 10 Commandments Transforming Brand Culture in a Consumer Democracy* (New York: Allworth Press, 2002) and Michael Willmott *Citizen Brands: Putting Society at the Heart of Your Business* (New York: Wiley, 2002).
74. Ouellette, "Citizen Brand," 69–70.
75. For more on slacktivism, see Evgeny Morozov, *The Net Delusion: The Dark Side of Internet Freedom* (New York: Public Affairs, 2011), 179–204.
76. Arvidsson, "Ethical Economy," 17.
77. Banet-Weiser, *Authentic*, 9. Citing Raymond Williams, *The Long Revolution* (Peterborough, ON: Broadview Press, 1961), 64.
78. Roger Silverstone, *Television and Everyday Life* (New York: Routledge, 1994), 2.
79. Paddy Scannell, *Radio, Television, and Modern Life: A Phenomenological Approach* (Oxford: Blackwell, 1996), 23.
80. Scannell, *Radio, Television, and Modern Life*, 145.
81. Tussey, 37.
82. Henri Lefebvre, *Critique of Everyday Life, Volume 2: Foundations for a Sociology of the Everyday* (New York: Verso, 2002), 75–77; Lefebvre, *Rhythmanalysis: Space, Time, and Everyday Life* (London: Continuum, 2004), 46.
83. Lefebvre, *Rhythmanalysis*, 50.

CONCLUSION

1. Todd Spangler, "Twitter Pushes Live-Video Deals With MLB, NFL, Viacom, BuzzFeed, Live Nation, WNBA and More," *Variety*, May 1, 2017, http://variety.com/2017/digital/news/twitter-pushes-live-video-deals-with-mlb-buzzfeed-live-nation-wnba-and-others-1202405236/.

2. Spangler, "Twitter Push Live-Video Deals With MLB, NFL, Viacom, BuzzFeed, Live Nation, WNBA and More."

3. Spangler, "Twitter Push Live-Video Deals With MLB, NFL, Viacom, BuzzFeed, Live Nation, WNBA and More."

4. "Introducing Facebook Watch, a New Platform for Shows on Facebook," *Facebook Newsroom*, August 9, 2017, https://newsroom.fb.com/news/2017/08/introducing-watch-a-new-platform-for-shows-on-facebook/.

5. Sahil Patel, "Facebook Recruits Top Publishers for Exclusive Shows," *Digiday*, May 25, 2017, https://digiday.com/media/facebook-recruits-top-video-publishers-shows/.

6. Patel, "Facebook Starts Rolling out Funded Shows for Watch," *Digiday*, August 30, 2017, https://digiday.com/media/come-facebook-shows/.

7. Scott Soshnick, Sarah Frier, and Scott Moritz, "Twitter Gets NFL Thursday Night Games for a Bargain Price," *Bloomberg*, April 5, 2016, https://www.bloomberg.com/news/articles/2016-04-05/twitter-said-to-win-nfl-deal-for-thursday-night-streaming-rights.

8. Sarah Perez, "Twitter Announces a New Deal for Year-Round NFL Content That Includes Live Video, but No Games," *TechCrunch*, May 11, 2017, https://techcrunch.com/2017/05/11/twitter-announces-a-new-deal-for-year-round-nfl-content-that-includes-live-video-but-no-games/.

9. John Herrman, "Territory Annexed," *The Awl*, January 8, 2015, http://www.theawl.com/2015/01/everything-ends.

10. Cale Guthrie Weissman, "For Digital Publishers, the 'Pivot to Video' Bloodbath is Here," *Fast Company*, January 12, 2018, https://www.fastcompany.com/40516189/for-digital-publishers-the-pivot-to-video-bloodbath-is-here.

11. Kurt Wagner, "Facebook Wants You Streaming Live Video from Whatever Device You Choose," *Recode*, April 12, 2016, https://www.recode.net/2016/4/12/11586064/facebook-live-video-api-launch-f8.

12. For more about the rise of cable television, see Megan Mullen, *Television in the Multichannel Age: A Brief History of Cable Television* (Malden: Wiley-Blackwell, 2008).

13. Cable and satellite companies lost 3.2 million subscribers in 2018 and 5.5 million subscribers in 2019. That is a cord-cutting rate increase of 70 percent year-over-year. See Lillian Rizzo and Drew FitzGerald, "Cord-Cutting Accelerated in 2019, Raising Pressure on Cable Providers," *Wall Street Journal*, February 20, 2020, http://www.wsj.com/articles/cord-cutting-accelerates-raising-pressure-on-cable-providers-11582149209.

14. Rick Porter, "TV Long View: Five Years of Network Ratings Declines in Context," *Hollywood Reporter*, September 21, 2019, https://www.hollywoodreporter.com/live-feed/five-years-network-ratings-declines-explained-1241524.

15. Netflix subscriber base grew to 182 million globally in the first quarter of 2020. See Edmund Lee, "Everyone You Know Just Signed Up for Netflix," *New York Times*, April 21,

2020, http://www.nytimes.com/2020/04/21/business/media/netflix-q1-2020-earnings-nflx.html.

16. In 2019 and 2020 alone, new streaming portals from Apple, Disney, WarnerMedia, and Comcast have hit the marketplace, putting further pressure on conventional television distribution and viewing. See Joe Flint and David Marcelis, "Streaming War Spurs Classic-TV Arms Race," *Wall Street Journal*, September 17, 2019, https://www.wsj.com/articles/streaming-services-splurge-on-old-tv-shows-to-fend-off-rivals-11568764309.

17. Stuart Cunningham, and David Craig, *Social Media Entertainment: The New Intersection of Hollywood and Silicon Valley* (New York: NYU Press, 2019).

18. See, for instance, Todd Spangler, "'Tiger King' Nabbed Over 34 Million US Viewers in First 10 Days, Nielsen Says," *Variety*, April 8, 2020, https://variety.com/2020/digital/news/tiger-king-nielsen-viewership-data-stranger-things-1234573602/.

19. William Uricchio, "The Future of a Medium Once Known as Television," in *The YouTube Reader*, ed. Pelle Snickars and Patrick Vonderau (Stockholm: National Library of Sweden, 2009), 29.

20. Richard Grusin, "YouTube at the End of New Media," in *The YouTube Reader*, ed. Pelle Snickars and Patrick Vonderau (Stockholm: National Library of Sweden, 2009), 63.

21. Julia Alexander, "YouTube Premium is Changing Because It Has To," *The Verge*, November 29, 2018, https://www.theverge.com/2018/11/29/18116154/youtube-premium-free-ads-subscription-red.

22. Davey Alba, "YouTube TV Goes Live in Google's Biggest Swipe at Comcast Yet," *Wired*, April 5, 2017, https://www.wired.com/2017/04/youtube-tv-review/.

23. Tarleton Gillespie, "The Politics of 'Platforms,'" *New Media & Society* 12, no. 3 (2010): 349.

24. Gillespie, "The Politics of 'Platforms,'" 359.

25. Ethan Tussey, *The Procrastination Economy: The Big Business of Downtime* (New York: NYU Press, 2018), 37–59.

26. Tussey, *The Procrastination Economy*, 143.

27. Grusin, *Premediation: Affect and Mediality after 9/11* (Basingstoke: Palgrave Macmillan, 2010), 125.

28. Elizabeth Moor, "Branded Spaces: The Scope of 'New Marketing,'" *Journal of Consumer Culture* 3, no. 1 (2003): 39–60.

29. B. Joseph Pine and James H. Gilmore, *The Experience Economy* (Boston: Harvard Business School Press, 1999); Avi Shankar and Brett Horton, "Ambient Media: Advertising's New Media Opportunity," *International Journal of Advertising* 18, no. 3 (1999): 305–22.

30. Anna McCarthy, *Ambient Television: Visual Culture and Public Space* (Durham: Duke University Press, 2001).

31. Michel de Certeau, *The Practice of Everyday Life*, trans. Steven Rendall (Berkeley: University of California Press, 1984), 91–110.

32. Lev Manovich, "The Practice of Everyday (Media) Life," in *Video Vortex Reader: Responses to YouTube*, eds. Geert Lovink and Sabine Niederer (Rotterdam: Institute of Network Cultures, 2008), 36–38.

33. Henri Lefebvre, *Critique of Everyday Life, Volume 2: Foundations for a Sociology of the Everyday* (London and New York: Verso, 2002), 75–77.

BIBLIOGRAPHY

Acland, Charles R. *Screen Traffic: Movies, Multiplexes, and Global Culture*. Durham: Duke University Press, 2003.

AlMarshedi, Alaa, Gary Wills, and Ashok Ranchhod. "Gamifying Self-Management of Chronic Illnesses: A Mixed-Methods Study." *JMIR Serious Games* 4, no. 2 (2016): n.p.

Anderson, Christopher. "Producing an Aristocracy of Culture in American Television." In *The Essential HBO Reader*, eds. Gary Edgerton and Jeffrey P. Jones, 23–41. Lexington: University Press of Kentucky, 2008. 23–41.

Anderson, Steve F. *Technologies of History: Visual Media and the Eccentricity of the Past*. Hanover: Dartmouth College Press, 2011.

Andrejevic, Mark. "Exploitation in the Data Mine." In *Internet and Surveillance: The Challenges of Web 2.0 and Social Media*, eds. Christian Fuchs, Kees Boersma, Anders Albrechtslund, and Marisol Sandoval, 71–88. New York: Routledge, 2012.

Andrejevic, Mark. *iSpy: Surveillance and Power in the Interactive Era*. Lawrence: University Press of Kansas, 2007.

Ang, Ien. *Desperately Seeking the Audience*. London: Routledge, 1991.

Aronczyk, Melissa. "Portal or Police? The Limits of Promotional Paratexts." *Critical Studies in Media Communication* 34, no. 2 (2017): 111–19.

Arvidsson, Adam. *Brands: Meaning and Value in Media Culture*. New York: Routledge, 2006.

Arvidsson, Adam. "The Ethical Economy: Towards a Post-Capitalist Theory of Value." *Capital and Class*, no. 97 (2009): 13–29.

Auslander, Philip. *Liveness: Performance in a Mediatized Culture, Second Edition*. London: Routledge, 2008.

Banet-Weiser, Sarah. *Authentic (TM): The Politics of Ambivalence in a Brand Culture*. New York: NYU Press, 2012.

Banks, John, and Mark Deuze. "Co-Creative Labor." *International Journal of Cultural Studies* 12, no. 5 (2009): 419–31.

Banks, John, and Sal Humphreys. "The Labor of User Co-Creators: Emergent Social Network Markets?" *Convergence* 14, no. 4 (2008): 401–18.

Barbrook, Richard. "The Hi-Tech Gift Economy." *First Monday* 3, no. 12 (1998/2005). http://firstmonday.org/htbin/cgiwrap/bin/ojs/index.php/fm/article/view/631/552.

Barker, Cory. "'Social TV': *Pretty Little Liars*, Casual Fandom, Celebrity Instagramming, and Media Life." *Journal of Popular Culture Studies* 2, nos. 1–2 (2014): 215–42.

Beniger, James R. *The Control Revolution*. Cambridge: Harvard University Press, 1986.

Bjarkman, Kim. "To Have and to Hold: The Video Collector's Relationship to an Ethereal Medium." *Television & New Media* 5, no. 3 (2004): 217–46.

Boddy, William. *New Media and Popular Imagination*. Oxford: Oxford University Press, 2004.

Bolter Jay David, and Richard Grusin. *Remediation: Understanding New Media*. Cambridge: MIT Press, 1999.

Boni, Marta, ed. *World Building: Transmedia, Fans, Industries*. Amsterdam: Amsterdam University Press, 2017.

Booth, Paul. *Digital Fandom: New Media Studies*. New York: Peter Lang, 2010.

Booth, Paul. *Negotiating Fandom and Media in the Digital Age*. Iowa City: University of Iowa Press, 2015.

Bourdon, Jérôme. "Live Television Is Still Alive: On Television as An Unfulfilled Promised." *Media, Culture & Society*, no. 22 (2000): 531–56.

boyd, danah. "Social Media is Here to Stay . . . Now What?" Microsoft Research Tech Fest, Redmond, WA, February 26, 2009. http://www.danah.org/papers/talks/MSRTechFest2009.html.

Brabham, Daren C. "The Myth of Amateur Crowds: A Critical Discourse Analysis of Crowdsourcing Coverage." *Information, Communication & Society* 15, no. 3 (2012): 394–410.

Brabham, Daren C. "Crowdsourcing as a Model for Problem Solving: An Introduction and Cases." *Convergence* 14, no. 1 (2008): 75–90.

Branston, Gill. "Histories of British Television." In *The Television Studies Book*, eds. Christine Geraghty and David Lusted, 51–62. London: Arnold, 1998.

Brock Jr., André. *Distributed Blackness: African American Cybercultures*. New York: NYU Press, 2020.

Broll, Wolfgang, Jan Ohlenburg, Irma Lindt, Iris Herbst, and Anne-Kathrin Braun. "Meeting Technology Challenges of Pervasive Augmented Reality Games." Presented at the *Proceedings of the 5th ACM SIGCOMM Workshop on Network and Support System for Games—NetGames '06*, October 30–31, 2006. Article 28, n.p.

Brooker, Will. "Going Pro: Gendered Responses to the Incorporation of Fan Labor as User-Generated Content." In *Wired TV: Laboring Over an Interactive Future*, ed. Denise Mann, 72–97. New Brunswick: Rutgers University Press, 2014.

Brooker, Will. "Living on *Dawson's Creek*: Teen Viewers, Cultural Convergence, and Television Overflow." *International Journal of Cultural Studies* 4, no. 4 (2001): 456–72.

Brooker, Will. "Television Out of Time: Watching Cult Shows on Download." In *Reading Lost: Perspectives on a Hit Television Show*, ed. Roberta Pearson, 51–72. London: I. B. Tauris, 2009.

Brookey, Robert, and Jonathan Gray. "Not Merely Para: Continuing Steps in Paratextual Research." *Critical Studies in Media Communication* 34, no. 2 (2017): 101–10.

Brookey, Robert, and Robert Westerfelhaus. "Hiding Homoeroticism in Plain View: The *Fight Club* DVD as Digital Closet." *Critical Studies in Media Communication* 19, no. 1 (2002): 21–43.

Bruns, Axel. *Blogs, Wikipedia, Second Life, and Beyond: From Production to Produsage*. New York: Peter Lang, 2008.

Burgess, Jean, and Joshua Green. *YouTube: Online Video and Participatory Culture*. Cambridge: Polity Press, 2009.

Busse, Kristina. "Geek Hierarchies, Boundary Policing, and the Gendering of the Good Fan." *Participations* 10, no. 1 (2013): 73–91.

Caldwell, John T. "Convergence Television: Aggregating Form and Repurposing Content in the Culture of Conglomeration." In *Television After TV: Essays on a Medium in Transition*, eds. Lynn Spigel and Jan Olsson, 41–74. Durham: Duke University Press, 2004.

Caldwell, John T. "Prefiguring DVD Bonus Tracks: Making-ofs and Behind-the-Scenes as Historic Television Programming Strategies Prototypes." In *Film and Television After DVD*, eds. James Bennett and Tom Brown, 149–71. London: Routledge, 2008.

Caldwell, John T. *Production Culture: Industrial Reflexivity and Critical Practice in Film and Television*. Durham: Duke University Press, 2008.

Caldwell, John T. "'Second-Shift' Media Aesthetics: Programming, Interactivity, and User Flows." In *New Media: Theories and Practices of Digitextuality*, eds. John T. Caldwell and Anna Everett, 127–44. London: Routledge, 2003.

Caldwell, Sarah. "Is Quality TV Any Good? Generic Distinctions, Evaluations and the Troubling Matter of Critical Judgment." In *Quality TV: Contemporary American Television and Beyond*, eds. Janet McCabe and Kim Akass, 19–34. New York: I. B. Tauris, 2007.

Carah, Nicholas. "Algorithmic Brands: A Decade of Brand Experiments with Mobile and Social Media." *New Media & Society* 19, no. 3 (2017): 384–400.

Carey, James, and John J. Quirk. "The Mythos of the Electronic Revolution." In *Communication as Culture: Essays on Media and Society*, ed. James Carey, 113–41. New York: Routledge, 1989.

Carpentier, Nico. *Media and Participation: A Site of Ideological-Democratic Struggle*. Chicago: Intellect, 2011.

Chamberlin, Daniel. "Media Interfaces, Networked Media Spaces, and the Mass Customization of Everyday Space." In *Flow TV: Television in the Age of Modern Convergence*, eds. Michael Kackman, Marnie Binfield, Matthew Thomas Payne, Allison Perlman, and Bryan Sebok, 13–29. New York: Routledge, 2011.

Christensen, Clayton. *The Innovator's Dilemma: When New Technologies Cause Great Firms to Fail*. New York: Harvard Business Review Press, 1997.

Coffey, Amy Jo. "Promotional Practices of Cable Networks: A Comparative Analysis of New and Traditional Spaces." *International Journal on Media Management* 13, no. 3 (2011): 161–76.

Couldry, Nick. "Liveness, 'Reality,' and the Mediated Habitus from Television to the Mobile Phone." *Communication Review* 7, no. 4 (2004): 353–61.

Couldry, Nick. *Media Rituals: A Critical Approach*. London: Routledge, 2003.

Cunningham, Stuart, and David Craig. *Social Media Entertainment: The New Intersection of Hollywood and Silicon Valley*. New York NYU Press, 2019.

Danan, Martine. "Marketing the Hollywood Blockbuster in France." *Journal of Popular Film and Television* 23, no. 2 (1995): 131–40.

Dawson, Max. "Little Players, Big Shows: Format, Narrative, and Style on Television's New Smaller Screens." *Convergence* 13, no. 3 (2007): 231–50.

Dawson, Max. "Television Abridged: Ephemeral Texts, Monumental Seriality, and TV-Digital Media Convergence." In *Ephemeral Media: Transitory Screen Culture from Television to YouTube*, ed. Paul Grainge, 37–58. London: British Film Institute, 2011.

Dawson, Max. "Television's Aesthetic of Efficiency: Convergence Television and the Digital Short." In *Television as Digital Media*, eds. James Bennett and Niki Strange, 204–29. Durham: Duke University Press, 2011.

Dawson, Max. *TV Repair: New Media 'Solutions' to Old Media Problems*. PhD dissertation, Northwestern University, 2008.

Dayan, Daniel, and Elihu Katz. *Media Events: The Live Broadcasting of History*. Cambridge: Harvard University Press, 1994.

de Certeau, Michel. *The Practice of Everyday Life*: Berkley: University of California Press, 1984.

DeFino, Dean J. *The HBO Effect*. New York: Bloomsbury Academic, 2013.

De Kosnik, Abigail. "Should Fan Fiction Be Free?" *Cinema Journal* 48, no. 4 (2009): 118–24.

Deller, Ruth. "Twittering On: Audience Research and Participation Using Twitter." *Participations* 8, no. 1 (2011): 216–45.

Derhy Kurtz, Benjamin W. L., and Mélanie Bourdaa, eds. *The Rise of Transtexts: Challenges and Opportunities*. New York: Routledge, 2016.

Doyle, Gillian. "From Television to Multi-Platform: Less from More or More for Less?" *Convergence* 16, no. 4 (2010): 431–49.

Edgerton, Gary R., and Jeffrey P. Jones, eds. *The Essential HBO Reader*. Lexington: University Press of Kentucky, 2009.

Ellis, John. "Scheduling: The Last Creative Act in Television?" *Media, Culture & Society* 22, no. 1 (2000): 25–38.

Ellis, John. *Visible Fictions: Cinema: Television: Video*. New York: Routledge, 1982.

Eppink, Jason. "A Brief History of The GIF (So Far)." *Journal of Visual Culture* 13, no. 3 (2014): 298–306.

Erigha, Maryann. "Shonda Rhimes, *Scandal*, and the Politics of Crossing Over." *Black Scholar* 45, no. 1 (2015): 10–15.

Evans, Elizabeth. *Transmedia Television: Audiences, New Media, and Daily Life*. New York: Routledge, 2011.

Everett, Anna. "Scandalicious: *Scandal*, Social Media, and Shonda Rhimes's Auteurist Juggernaut." *Black Scholar* 45, no. 1 (2015): 34–43.

Feuer, Jane. "HBO and the Concept of Quality TV." In *Quality TV: Contemporary American Television and Beyond*, eds. Janet McCabe and Kim Akass, 145–57. New York: I. B. Tauris, 2007.

Feuer, Jane. "Live Television: The Concept of Ontology as Ideology." In *Regarding Television: Critical Approaches*, ed. E. Ann Kaplan, 12–22. New York: University Publications of America, 1983.

Fiske, John. "The Cultural Economy of Fandom." In *The Adoring Audience: Fan Culture and Popular Media*, ed. Lisa A. Lewis, 30–49. New York: Routledge, 1992.

Florini, Sarah. *Beyond Hashtags: Racial Politics and Black Digital Networks*. New York: NYU Press, 2019.

Florini, Sarah. "Tweets, Tweeps, and Signifyin': Communication and Cultural Performance on 'Black Twitter.'" *Television & New Media* 15, no. 3 (2014): 223–37.
Fuchs, Christian. "Labor in Informational Capitalism and On the Internet." *Information Society* 26, no. 3 (2010): 179–96.
Galloway, Alexander. *The Interface Effect*. New York: Polity Press, 2012.
Genette, Gerard. *Paratexts: The Thresholds of Interpretation*, translated by Jane E. Lewin. Cambridge: Cambridge University Press, 1997.
Gerlitz, Carolin, and Anne Helmond. "The Like Economy: Social Buttons and the Data-Intensive Web." *New Media & Society* 15, no. 8 (2013): 1348–65.
Gilder, George. *Life After Television: The Coming Transformation of Media and American Life*. New York: Norton, 1994.
Gillan, Jennifer. *Television and New Media: Must-Click TV*. New York: Routledge, 2011.
Gillan, Jennifer. *Television Brandcasting: The Return of the Content-Promotion Hybrid*. New York: Routledge, 2014.
Gillespie, Tarleton. "The Politics of Platforms." *New Media & Society* 12, no. 3 (2010): 347–64.
Gillespie, Tarleton. "The Relevance of Algorithms." In *Media Technologies: Essays on Communication, Materiality, and Society*, eds. Tarleton Gillespie, Pablo J. Boczkowski, and Kirsten A. Foot, 167–95. Cambridge: MIT Press, 2013.
Gitelman, Lisa. *Always Already New: Media History, and the Data of Culture*. Cambridge: MIT Press, 2008.
Gitlin, Todd. *Inside Prime Time*. New York: Pantheon, 1983.
Gobé, Mark. *Citizen Brand: 10 Commandments Transforming Brand Culture in a Consumer Democracy*. New York: Allworth Press, 2002.
Grainge, Paul. *Brand Hollywood: Selling Entertainment in a Global Media Age*. London: Routledge, 2007.
Grainge, Paul. "Introduction: Ephemeral Media." In *Ephemeral Media: Transitory Screen Culture from Television to YouTube*, ed. Paul Grainge, 1–19. London: British Film Institute, 2011.
Gray, Herman. *Watching Race: Television and the Struggle for Blackness*. Minneapolis: University of Minnesota Press, 2004.
Gray, Jonathan. "Reviving Audience Studies." *Critical Studies in Media Communication* 34, no. 1 (2017): 79–83.
Gray, Jonathan. *Show Sold Separately: Promos, Spoilers, and Other Media Paratexts*. New York: NYU Press, 2010.
Green, Joshua. "Why Do They Call It TV When It's Not on The Box? 'New' Television Services and Old Television Functions." *Media International Australia*, no. 126 (2008): 95–105.
Greer, Clark F., and Douglas A. Ferguson. "Using Twitter for Promotion and Branding: A Content Analysis of Local Television Twitter Sites." *Journal of Broadcasting & Electronic Media* 55, no. 2 (2011): 198–214.
Grusin, Richard. *Premediation: Affect and Mediality after 9/11*. Basingstoke: Palgrave Macmillan, 2010.

Hadas, Leora. "A New Vision: J. J. Abrams, *Star Trek*, and Promotional Authorship." *Cinema Journal* 56, no. 2 (2017): 46–66.

Halavais, Alexander M. C. "A Genealogy of Badges: Inherited Meaning and Monstrous Moral Hybrids." *Information, Communication & Society* 15, no. 3 (2012): 354–73.

Hamari, Juho, and Jonna Koivisto. "'Working Out for Likes': An Empirical Study on Social Influence in Exercise Gamification." *Computers in Human Behavior*, no. 50 (2015): 333–47.

Hardt, Michael. "Affective Labor." *Boundaries*, no. 2 (1999): 89–100.

Harrington, Stephen, Tim Highfield, and Axel Bruns. "More than a Backchannel: Twitter and Television." *Participations* 10, no. 1 (2013): 405–9.

Hastie, Amelie. "Detritus and the Moving Image: Ephemera, Materiality, History." *Journal of Visual Culture* 6, no. 2 (2007): 171–74.

Hellekson, Karen. "A Fannish Field of Value: Online Gift Culture." *Cinema Journal* 48, no. 4 (2009): 113–18.

Hessler, Jennifer. "Peoplemeter Technologies and the Biometric Turn in Audience Measurement." *Television & New Media* 22, no. 4 (2021): 400–419.

Hight, Craig. "Making-of Documentaries on DVD: *The Lord of the Rings* Trilogy and Special Editions." *Velvet Light Trap*, no. 56 (2005): 4–17.

Hills, Matt. "From the Box in the Corner to the Box Set on the Shelf: 'TVIII' and the Cultural/Textual Valorizations of DVD." *New Review of Film and Television Studies* 5, no. 1 (2007): 41–60.

Hills, Matt. "*Veronica Mars*, Fandom, and the 'Affective Economics' of Crowdfunding." *New Media & Society* 17, no. 2 (2015): 183–97.

Hilmes, Michele. *Only Connect: A Cultural History of Broadcasting in the United States*. Belmont: Wadsworth, 2002.

Holt, Jennifer, and Kevin Sanson, eds. *Connected Viewing: Selling, Streaming, and Sharing Media in the Digital Age*. New York: Routledge, 2013.

Hughes, James. *Citizen Cyborg: Why Democratic Societies Must Respond to the Redesigned Human of the Future*. Cambridge: Basic Books, 2004.

Hyde, Lewis. *The Gift: Creativity and the Artist in the Modern World, 25th Anniversary Edition*. New York: Vintage Books, 2009.

Iampolski, Mikhail. *The Memory of Tiresias: Intertextuality and Film*, translated by Harsha Ram. Berkeley: University of California Press, 1998.

Ihlebaek, Karoline Andrea, Trine Syvertsen, and Espen Ytreberg. "Keeping Them and Moving Them: TV Scheduling in the Phase of Channel and Platform Proliferation." *Television & New Media* 15, no. 5 (2014): 470–86.

Ingram-Waters, Mary, and Leslie Balderas. "Blurring Production Boundaries within Fan Empowerment." In *Adventures in Shondaland: Identity Politics and the Power of Representation*, eds. Michaela D. E. Meyer and Rachel Alicia Griffin, 197–213. New Brunswick: Rutgers University Press, 2018.

Jaramillo, Deborah L. "AMC: Stumbling Toward a New Television Canon." *Television & New Media* 14, no. 2 (2012): 167–83.

Jaramillo, Deborah L. "The Family Racket: AOL-Time Warner, HBO, *The Sopranos*, and the Construction of a Quality Brand." *Journal of Communication Inquiry* 26, no. 1 (2002): 59–75.

Jenkins, Henry. *Convergence Culture: When Old and New Media Collide*. New York: NYU Press, 2006.

Jenkins, Henry. *Fans, Bloggers, and Gamers: Exploring Participatory Culture*. New York: NYU Press, 2006.

Jenkins, Henry. "'We Had So Much Story to Tell': The *Heroes* Comics as Transmedia Storytelling." *Confessions of An Aca-Fan*, December 3, 2007. http://henryjenkins.org/blog/2007/12/we-had-so-many-stories-to-tell.html.

Jenkins, Henry, and Nico Carpentier. "Theorizing Participatory Intensities: A Conversation about Participation and Politics." *Convergence* 19, no. 3 (2013): 265–86.

Jenkins, Henry, Sam Ford, and Joshua Green. *Spreadable Media: Creating Meaning and Value in a Networked Culture*. New York: NYU Press, 2013.

Johnson, Catherine. *Branding Television*. New York: Routledge, 2012.

Johnson, Catherine. "Tele-Branding in TVIII: The Network as Brand and the Programme as Brand." *New Review of Film and Television Studies* 5, no. 1 (2007): 5–24.

Johnson, Derek. "Devaluing and Revaluing Seriality: The Gendered Discourses of Media Franchising." *Media, Culture, and Society* 33, no. 7 (2011): 1077–93.

Johnson, Derek. "Inviting Audiences In: The Spatial Organization of Production and Consumption." *New Review of Film and Television Studies* 5, no. 1 (2007): 61–80.

Johnson, Richard. "Historical Returns: Transdisciplinarity, Cultural Studies, and History." *European Journal of Cultural Studies* 4, no. 3 (2001): 261–88.

Joseph, Ralina L. "Strategically Ambiguous Shonda Rhimes: Respectability Politics of a Black Woman Showrunner." *Souls* 18, nos. 2–4 (2016): 302–20.

Klein, Naomi. *No Logo: Taking Aim at the Brand Bullies*. New York: Picador, 2002.

Klinger, Barbara. *Beyond the Multiplex: Cinema, New Technologies, and the Home*. Berkley: University of California Press, 2006.

Kohnen, Melanie E. S. "Cultural Diversity as Brand Management in Cable Television." *Media Industries Journal* 2, no. 2 (2015). https://doi.org/10.3998/mij.15031809.0002.205.

Kompare, Derek. "Publishing Flow: DVD Box Sets and the Reconception of Television." *Television & New Media* 7, no. 4 (2006): 335–60.

Kristeva, Julia. *Desire in Language: A Semiotic Approach to Literature and Art*, trans. Thomas Gora, ed. Leon Roudiez. Oxford: Blackwell, 1980.

Lazzarato, Maurizio. "Immaterial Labor." In *Radical Thought in Italy: A Potential Politics*, eds. Paolo Virno and Michael Hardt, 133–50. Minneapolis: University of Minnesota Press, 1996.

Lee, Hye Jin, and Mark Andrejevic. "Second-Screen Theory: From the Democratic Surround to the Digital Enclosure." In *Connected Viewing: Selling, Streaming, & Sharing Media in the Digital Era*, eds. Jennifer Holt and Kevin Sanson, 40–61. New York: Routledge, 2014.

Lefebvre, Henri. *Critique of Everyday Life, Volume 2: Foundations for a Sociology of the Everyday*. London and New York: Verso, 2002.

Lefebvre, Henri. *Rhythmanalysis: Space, Time, and Everyday Life*. London: Continuum, 2004.

Levine, Elana. "Distinguishing Television: The Changing Meanings of Television Liveness." *Media, Culture & Society* 30, no. 3 (2008): 393–409.

Lévy, Pierre. *Collective Intelligence: Mankind's Emerging World in Cyberspace*, translated by Robert Bononno. New York: Plenum, 1997.

Li, Xiaochang. "Fanfic, Inc.: Another Look at Fanlib.com." *MIT Convergence Culture Consortium C3 Weekly Update*, December 7, 2007. http://www.convergenceculture.org/htmlnewsletter/weeklyupdate_20071207.html.

Li, Xiaochang. *More than Money Can Buy: Locating Value in Spreadable Media*. Cambridge: MIT Convergence Culture Consortium, 2009. http://convergenceculture.org/research/C3LocatingValueWhitePaper.pdf.

Light, Jennifer S. "Before the Internet, There Was Cable." *IEEE Annals of the History of Computing* 25, no. 2 (2003): 94–96.

Lister, Cameron, Joshua H. West, Ben Cannon, Tyler Sax, and David Brodegard. "Just a Fad? Gamification in Health and Fitness Apps." *JMIR Serious Games* 2, no. 2 (2014): n.p.

Livingstone, Sonia. "Audiences in the Age of Datafication: Critical Questions for Media Research." *Television & New Media* 20, no. 2 (2019): 170–83.

Lotz, Amanda D. *Cable Guys: Television and Masculinities in the 21st Century*. New York: NYU Press, 2014.

Lotz, Amanda D. "If It's Not TV, What Is It? The Case of U.S. Subscription Television." In *Cable Visions: Television Beyond Broadcasting*, eds. Sarah Banet-Weiser, Cynthia Chris, and Anthony Freitas, 85–102. New York: NYU Press, 2007.

Lotz, Amanda D. *Portals: A Treatise on Internet-Distributed Television*. Ann Arbor: University of Michigan Publishing, 2017.

Lotz, Amanda D. *Redesigning Women: Television After the Network Era*. New York: Routledge, 2006.

Lovink, Geert. *Networks Without A Cause: A Critique of Social Media*. Cambridge: Polity Press, 2011.

Lury, Celia. "Brand as Assemblage: Assembling Culture." *Journal of Cultural Economy* 2, nos. 1–2 (2009): 67–82.

Lury, Celia. *Brands: The Logos of the Global Economy*. New York: Routledge, 2004.

Mann, Denise, ed. *Wired TV: Laboring Over an Interactive Future*. New Brunswick: Rutgers University Press, 2014.

Manovich, Lev. *The Language of New Media*. Cambridge: MIT Press, 1999.

Manovich, Lev. "The Practice of Everyday (Media) Life." In *Video Vortex Reader: Responses to YouTube*, eds. Geert Lovink and Sabine Niederer, 33–44. Rotterdam: Institute of Network Cultures, 2008.

McCabe, Janet, and Kim Akass. "Introduction: Debating Quality." In *Quality TV: Contemporary American Television and Beyond*, eds. Janet McCabe and Kim Akass, 1–12. New York: I. B. Tauris, 2007.

McCabe, Janet. "It's Not TV, It's HBO Original Programming: Producing Quality TV." In *It's Not TV: Watching HBO in the Post-Television Era*, eds. Marc Leverette, Brian L. Ott, and Cara Louise Buckley, 83–94. New York: Routledge, 2008.

McCarthy, Anna. *Ambient Television: Visual Culture and Public Space*. Durham: Duke University Press, 2001.

McDonald, Paul. *Video and DVD Industries*. London: British Film Institute, 2008.

McMurria, John. "Long-Format TV: Globalization and Network Branding in a Multi-Channel Era." In *Quality Popular Television: Cult TV, The Industry, and Fans*, eds. Mark Jancovich and James Lyons, 65–87. London: British Film Institute, 2003.

McNutt, Myles. "Tweets of Anarchy: Showrunners on Twitter." *Antenna*, September 17, 2010. http://blog.commarts.wisc.edu/2010/09/17/tweets-of-anarchy-showrunners-on-twitter/.

Meyer, Michaela D. E., and Rachel Alicia Griffin, eds. *Adventures in Shondaland: Identity Politics and the Power of Representation*. New Brunswick: Rutgers University Press, 2018.

Miller, Vincent. "New Media, Networking, and Phatic Culture." *Convergence* 14, no. 4 (2008): 387–400.

Milner, Ryan M. "Working for the Text: Fan Labor and the New Organization." *International Journal of Cultural Studies* 12, no. 5 (2009): 491–508.

Mittell, Jason. *Complex TV: The Poetics of Contemporary Television Storytelling*. New York: NYU Press, 2015.

Mittell, Jason. "Forensic Fandom and the Drillable Text." *Spreadable Media* (exclusive web essay). http://spreadablemedia.org/essays/mittell/.

Moor, Elizabeth. "Branded Spaces: The Scope of 'New Marketing.'" *Journal of Consumer Culture* 3, no. 1 (2003): 39–60.

Morozov, Evgeny. *The Net Delusion: The Dark Side of Internet Freedom*. New York: Public Affairs, 2011.

Mukherjee, Roopali, and Sarah Banet-Weiser, eds. *Commodity Activism: Cultural Resistance in Neoliberal Times*. New York: NYU Press, 2012.

Mullen, Megan. *The Rise of Cable Programming in the United States: Revolution or Evolution?* Austin: University of Texas Press, 2009.

Mullen, Megan. *Television in the Multichannel Age: A Brief History of Cable Television*. Malden: Wiley-Blackwell, 2008.

Murray, Simone. "'Celebrating the Story the Way It Is': Cultural Studies, Corporate Media, and the Contested Utility of Fandom." *Continuum* 18, no. 1 (2004): 7–25.

Negroponte, Nicolas. *Being Digital*. London: Hodder and Stoughton, 1995.

Nelson, Robin. "'Quality Television': 'The Sopranos Is the Best Television Drama Ever . . . in My Humble Opinion.'" *Critical Studies in Television* 1, no. 1 (2006): 58–71.

Newman, Michael Z. "GIFs: The Attainable Text." *Film Criticism* 40, no. 1 (2016). https://doi.org/10.3998/fc.13761232.0040.123.

Newman, Michael Z., and Elana Levine. *Legitimating Television: Media Convergence and Cultural Status*. New York: Routledge, 2011.

Nygaard, Taylor. "Girls Just Want to be 'Quality': Lena Dunham, and *Girls*' Conflicting Brand Identity." *Feminist Media Studies* 13, no. 2 (2013): 253–62.

Nygaard, Taylor, and Jorie Lagerwey. "Broadcasting Quality: Re-Centering Feminist Discourse with *The Good Wife*." *Television & New Media* 18, no. 2 (2017): 105–13.

Ott, Brian L. "Introduction: The Not TV Text." In *It's Not TV: Watching HBO in the Post-Television Era*, eds. Marc Leverette, Brian L. Ott, and Cara Louise Buckley, 95–100. New York: Routledge, 2008.

Ouellette, Laurie. "Citizen Brand: ABC and the Do Good Turn in US Television." In *Commodity Activism: Cultural Resistance in Neoliberal Times*, eds. Roopali Mukherjee and Sarah Banet-Weiser, 57–75. New York: NYU Press, 2012.

Packard, Vance. *The Hidden Persuaders*. New York: IG Publishing, 2007 [1957].

Papacharissi, Zizi. "Without You, I'm Nothing: Performances of the Self on Twitter." *International Journal of Communication*, no. 6 (2012): 1989–2006.

Parker, Mark, and Deborah Parker. *The DVD and the Study of Film: The Attainable Text.* New York: Palgrave Macmillan, 2011.

Perren, Alisa. *Indie, Inc.: Miramax and the Transformation of Hollywood in the Late 1990s.* Austin: University of Texas Press, 2012.

Perren, Alisa, and Jennifer Holt, eds. *Media Industries Studies: History, Theory, and Method.* New York: Wiley, 2011.

Pine, B. Joseph, and James H. Gilmore. *The Experience Economy.* Boston: Harvard Business School Press, 1999.

Pittman, Matthew, and Alec C. Tefertiller. "Without or Without You: Connected Viewing and Co-Viewing Twitter Activity for Traditional Appointment and Asynchronous Broadcast Television Models." *First Monday* 20, no. 7 (2015). http://firstmonday.org/ojs/index.php/fm/article/view/5935.

Polan, Dana. "Cable Watching: HBO, *The Sopranos*, and Discourses of Distinction." In *Cable Visions: Television Beyond Broadcasting*, eds. Sarah Banet-Weiser, Cynthia Chris, and Anthony Freita, 261–83. New York: NYU Press, 2007.

Proulx, Mike, and Stacey Shepatin. *Social TV: How Marketers Can Research and Engage Audiences by Connecting Television to the Web, Social Media, and Mobile.* Hoboken: Wiley, 2012.

Rheingold, Howard. *The Virtual Community: Homesteading on the Electronic Frontier.* Malden: MIT Press, 1993.

Riffaterre, Michael. "Compulsory Reader Response: The Intertextual Drive." In *Intertextuality: Theories and Practices*, eds. Michael Worton and Judith Still, 56–78. Manchester: Manchester University Press, 1990.

Rojek, Chris. *Capitalism and Leisure Theory.* London: Tavistock, 1985.

Ross, Sharon Marie. *Beyond the Box: Television and the Internet.* Malden: Wiley, 2008.

Salter, Anastasia, and Mel Stanfill. *A Portrait of the Auteur as Fanboy: The Construction of Authorship in Transmedia Franchises.* Jackson: University Press of Mississippi, 2020.

Sandler, Kevin. "'A Kid's Gotta Do What a Kid's Gotta Do': Branding the Nickelodeon Experience." In *Nickelodeon Nation: The History, Politics, and Economics of America's Only TV Channel for Kids*, ed. Heather Hendershot, 62–99. New York: NYU Press, 2004.

Sandvoss, Cornel. "Fans Online: Affective Media Consumption and Production in the Age of Convergence." In *Online Territories: Globalization, Mediated Practice, and Social Space*, eds. Miyase Christensen, André Jansson, and Christian Christensen, 49–74. New York: Peter Lang, 2011.

Santo, Avi. "Para-Television and Discourses of Distinction: The Culture of Production at HBO." In *It's Not TV: Watching HBO in the Post-Television Era*, eds. Marc Leverette, Brian L. Ott, and Cara Louise Buckley, 19–43. New York: Routledge, 2008.

Scannell, Paddy. *Radio, Television, and Modern Life: A Phenomenological Approach.* Cambridge and Oxford: Blackwell, 1996.

Scannell, Paddy. *Television and the Meaning of "Live": An Inquiry into the Human Situation.* London: Polity Press, 2013.

Schmidt, Lisa. "Monstrous Melodrama: Expanding the Scope of Melodramatic Identification to Interpret Negative Fan Responses to *Supernatural*." *Transformative Works and Cultures*, no. 4 (2010). https://doi.org/10.3983/twc.2010.0152.

Schneider, Steven M., and Kirsten Foot. "The Web as an Object of Study." *New Media & Society* 6, no. 1 (2004): 114–22.

Sconce, Jeffrey. "What If? Charting Television's New Textual Boundaries." In *Television After TV: Essays on a Medium in Transition*, eds. Lynn Spigel and Jan Olsson, 93–112. Durham: Duke University Press, 2004.

Scott, Suzanne. "'Authorized Resistance': Is Fan Production Frakked?" In *Cylons in America: Critical Studies in Battlestar Galactica*, eds. Tiffany Potter and C. W. Marshall, 210–23. New York, Continuum, 2008.

Scott, Suzanne. *Fake Geek Girls: Fandom, Gender, and the Convergence Culture Industry*. New York: NYU Press, 2019.

Scott, Suzanne. "Repackaging Fan Culture: The Regifting Economy of Ancillary Content Models." *Transformative Works and Cultures*, no. 3 (2009). https://doi.org/10.3983/twc.2009.0150.

Scott, Suzanne. "Who's Steering the Mothership? The Role of the Fanboy Auteur in Transmedia Storytelling." In *The Participatory Cultures Handbook*, eds. Aaron Delwiche and Jennifer Jacobs Henderson, 43–53. New York: Routledge, 2013.

Selznick, Barbara. "Branding the Future: Syfy in the Post-Network Era." *Science Fiction Film and Television* 2, no. 2 (2009): 177–204.

Shankar, Avi, and Brett Horton. "Ambient Media: Advertising's New Media Opportunity." *International Journal of Advertising* 18, no. 3 (1999): 305–22.

Silverstone, Roger. *Television and Everyday Life*. London and New York: Routledge, 1994.

Smith, Anthony N. "Pursuing 'Generation Snowflake': *Mr. Robot* and the USA Network's Mission for Millennials." *Television & New Media* 20, no. 5 (2019): 443–59.

Smith, Anthony N. "Putting the Premium into Basic: Slow-Burn Narratives and the Loss-Leader Function of AMC's Original Drama Series." *Television & New Media* 14, no. 2 (2012): 150–66.

Smith, Greg M. *On A Silver Platter: CD-ROMs and The Promises of a New Technology*. New York: NYU Press, 1999.

Smith, Jo T. "DVD Technologies and the Art of Control." In *Film and Television After DVD*, eds. James Bennett and Tom Brown, 129–48. London: Routledge, 2008.

Smith, Ralph Lee. *The Wire Nation: Cable TV, The Electronic Communications Highway*. New York: Harper & Row, 1972.

Smith-Shomade, Beretta E. *Shaded Lives: African-American and Television*. New Brunswick: Rutgers University Press, 2002.

Spigel, Lynn. *Make Room for TV: Television and the Family Ideal in Postwar America*. Chicago: University of Chicago Press, 1992.

Spigel, Lynn, and Max Dawson. "Television and Digital Media." In *American Thought and Culture in the 21st Century*, eds. Catherine Morley and Martin Halliwell, 275–90. Edinburgh: Edinburgh University Press, 2008.

Stanfill, Mel. *Exploiting Fandom: How the Media Industry Seeks to Manipulate Fans*. Iowa City: University of Iowa Press, 2019.

Straw, Will. "The Circulatory Turn." In *The Wireless Spectrum: The Politics, Practices, and Poetics of Mobile Media*, eds. Barbara Crow, Michael Longford, and Kim Sawchuk, 17–28. Toronto: University of Toronto Press, 2010.

Striphas, Ted. "Algorithmic Culture." *European Journal of Cultural Studies* 18, nos. 4–5 (2015): 395–412.

Swan, Phillip. *TVDotCom: The Future of Interactive Television*. New York: TV Books, 2000.

Terranova, Tiziana. "Free Labor: Producing Culture for the Digital Economy." *Social Text* 18, no. 2 (2004): 33–58.

Thompson, Robert J. "Preface." In *Quality TV: Contemporary American Television and Beyond*, eds. Janet McCabe and Kim Akass, xvii–xx. New York: I. B. Tauris, 2007.

Thornton, Sarah. *Club Cultures: Music, Media, and Subcultural Capital*. Middletown: Wesleyan University Press, 1996.

Tomlinson, John. *The Culture of Speed: The Coming of Immediacy*. London: Sage, 2007.

Tryon, Chuck. *On-Demand Culture: Digital Delivery and the Future of Movies*. New Brunswick: Rutgers University Press, 2013.

Turner, Graeme. "Approaching the Cultures of Use: Netflix, Disruption, and the Audience." *Critical Studies in Television* 14, no. 2 (2019): 222–32.

Turner, Graeme. "'Liveness' and 'Sharedness' Outside the Box." *Flow*, April 8, 2011. http://www.flowjournal.org/2011/04/liveness-and-sharedness-outside-the-box/.

Tussey, Ethan. *The Procrastination Economy: The Big Business of Downtime*. New York: NYU Press, 2018.

Uricchio, William. "The Future of a Medium Once Known as Television." In *The YouTube Reader*, eds. Pelle Snickars and Patrick Vonderau, 24–39. Stockholm: National Library of Sweden, 2009.

Uricchio, William. "Television's Next Great Generation: Technology/Interface Culture/Flow." In *Television After TV: Essays on a Medium in Transition*, eds. Lynn Spigel and Jan Olsson, 163–83. Durham: Duke University Press, 2005.

van Dijck, José. *The Culture of Connectivity: A Critical History of Social Media*. Oxford: Oxford University Press, 2013.

van Dijck, José. "Users Like You? Theorizing Agency in User-Generated Content." *Media, Culture & Society* 31, no. 1 (2009): 41–58.

van Dijck, José, and David Nieborg. "Wikinomics and Its Discontents: A Critical Analysis of Web 2.0 Business Manifestos." *New Media & Society* 11, no. 5 (2009): 855–74.

van Es, Karin. "Social TV and the Participation Dilemma in NBC's *The Voice*." *Television & New Media* 17, no. 2 (2016): 108–23.

Vianello, Robert. "The Power Politics of 'Live' Television." *Journal of Film and Video* 37, no. 3 (1985): 26–40.

Walters, James. "Repeat Viewings: Television Analysis in the DVD Age." In *Film and Television After DVD*, eds. James Bennett and Tom Brown, 63–80. London: Routledge, 2008.

Wanzo, Rebecca. "African American Acafandom and Other Strangers: New Genealogies of Fan Studies." *Transformative Works and Cultures*, no. 20 (2015). https://doi.org/10.3983/twc.2015.0699.

Ward, Sam. "Investigating the Practices of Television Branding: An Afterword on Methodology." *Networking Knowledge* 7, no. 1 (2014): 55–63.

Warner, Kristen. "ABC's *Scandal* and Black Women's Fandom." In *Cupcakes, Pinterest, and Ladyporn: Feminized Popular Culture in the Early Twenty-First Century*, ed. Elana Levine, 32–50. Urbana: University of Illinois Press, 2015.

Warner, Kristen. *The Cultural Politics of Colorblind TV Casting*. New York: Routledge, 2015.
Warner, Kristen. "If Loving Olitz Is Wrong, I Don't Want to Be Right: ABC's *Scandal* and the Affect of Black Female Desire." *Black Scholar* 45, no. 1 (2015): 16–20.
Warner, Kristen. "The Racial Logic of *Grey's Anatomy*: Shonda Rhimes and Her 'Post-Civil Rights, Post-Feminist' Series." *Television & New Media* 16, no. 7 (2015): 631–47.
Wexelblat, Alan. "An Auteur in the Age of the Internet: JMS, *Babylon 5*, and the Net." In *Hop on Pop: The Politics and Pleasures of Popular Culture*, eds. Henry Jenkins, Jane Shattuc, and Tara McPherson, 209–226. Durham: Duke University Press, 2002.
Wheatley, Helen, editor. *Re-Viewing Television Histories: Critical Issues in Television Historiography*. New York: I. B. Tauris, 2007.
Williams, Apryl, and Vanessa Gonlin. "I Got All My Sisters with Me (on Black Twitter): Second Screening of *How to Get Away with Murder* as a Discourse of Black Womanhood." *Information, Communication & Society* 20, no. 7 (2017): 984–1004.
Williams, Raymond. *The Long Revolution*. Peterborough, ON: Broadview Press, 1961.
Williams, Raymond. *Television: Technology and Cultural Form*. London: Routledge, 1990.
Willmott, Michael. *Citizen Brands: Putting Society at the Heart of Your Business*. New York: Wiley, 2003.
Wilson, Sherryl. "In the Living Room: Second Screens and TV Audiences." *Television & New Media* 17, no. 2 (2016): 174–91.
Wohn, D. Yvette, and Eun-Kyung Na. "Tweeting about TV: Sharing Television Experiences Via Social Media Message Streams." *First Monday* 16, no. 3 (2011). https://firstmonday.org/article/view/3368/2779.
Wood, Megan M., and Linda Baughman. "*Glee* Fandom and Twitter: Something New, or More of the Same Old Thing?" *Communication Studies* 63, no. 3 (2012): 328–44.
Wyatt, Justin. "The Formation of the Major Independent: Miramax, New Line, and the New Hollywood." In *Contemporary Hollywood Cinema*, eds. Steve Neale and Murray Smith, 74–90. London: Routledge, 1998.
Zook, Kristal Brent. *Color by Fox: The Fox Network and the Revolution in Black Television*. New York: Oxford University Press, 1999.
Zubiaga, Arkaitz, Damiano Spina, Raquel Martínez, and Victor Fresno. "Real-Time Classification of Twitter Trends." *Journal of the Association of Information Science & Technology* 66, no. 3 (2015): 462–73.

INDEX

ABC: and Academy Awards, 3–5; fansourc-ing of, 118; as platform, 184; scheduling of, 34–35, 37, 42; and Viggle, 112–13. See also *Grey's Anatomy*; *How to Get Away with Murder*; *Off the Map*; Rhimes, Shonda; *Scandal*; #TGIT: early campaign
ABC Studios, 56
Abraham (*The Walking Dead* character), 73, 82
Academy Awards, 3–5, 138
Acland, Charles, 148–49
acoustic footprint, 94
activism, 13, 22, 146, 170–71, 173–74, 185
Ad Age, 93
adaptation: in Amazon's Pilot Season, 126, 139; of *Preacher* for television, 58; of *Skam* for Facebook Watch, 177; of *The Lord of the Rings* for film, 118; of *The Walking Dead* for television, 72–74
advertising: on check-ins, 94, 108–14; in print, 26, 28; revenue for, 43, 55; on second-screen experiences, 70, 78–80; targeting with, 111–12; on television, 7, 9–10, 21, 26, 28, 43, 45–46, 52–53, 56; and upfront, 43, 176. See also promotion
affective branding, 14, 145, 149, 172–73, 182, 186
affective labor, 91–93
After, The (TV series), 127–28, 130, 137, 141
algorithm: and Amazon, 119, 129, 142; and digital video technologies, 34; and mobile applications, 67; and social platforms, 19, 42, 114, 177; and streaming video portals, 186

Allen, Woody, 138–40
Alpha House (TV series), 126–27
Alternative Reality Game (ARG), 10–11, 39, 65
Altman, Rick, 15
Altschuler, John, 151
Amazon (corporation): and algorithmic recommendations, 119, 126, 143; Channels program, 137; Jeff Bezos as CEO of, 119, 140; Kindle Fire, 120; as online market-place, 22, 108, 119, 146; Prime Instant Video, 57, 116, 123, 128, 137, 139; Prime subscription program, 116, 119–20, 125, 128, 137, 139, 141–43; star rating system of, 22, 116, 119, 127–28, 133, 137–38; Unbox, 120; Video (digital rental store), 11, 178. See Amazon Pilot Season; Amazon Studios; Price, Roy
Amazon Pilot Season: and auteurs, 134, 136, 139–40; and *Bosch*, 127–30, 133–34, 137, 141; development of, 22, 116–17; end of, 140–43; and ephemeral labor, 118–20, 128–29, 141; and fansourcing, 22, 117–19, 141–42; and feedback, 22, 121–30, 133, 135–43; fifth season of, 138–39; first season of, 123–27; fourth season of, 138–39; and Joey Soloway, 124, 127, 130, 134–36; press coverage of, 119, 122–25, 130–31, 133–36, 139–41; and quality TV discourses, 132–36; second season of, 127–31; and *The After*, 127–28, 130, 137, 141; third season of, 136–37; and *Transparent*, 119, 124, 127–28, 130, 133–37; website/interface of, 117, 121, 125–26, 128, 136. See Price, Roy

246   INDEX

Amazon Studios: and Amazon Channels, 137; and auteurs, 134, 136, 139–40; development of, 22, 116, 120–23; and fansourcing, 22, 117–19, 141–42; and film distribution, 138; and film festivals, 138; and Harvey Weinstein, 138, 140; and Jack Epps Jr., 121–22; and Joey Soloway, 124, 127, 130, 134–36; press coverage of, 119, 120–25, 130, 133–36, 138–41; and quality TV discourses, 132–36; and Spike Lee, 138; and Steven Soderbergh, 136–38; website/interface of, 117, 121, 125–26, 128, 136; and Woody Allen, 138–40. *See* Amazon Pilot Season; Price, Roy
ambient advertising, 186
AMC: and corporation synergy, 77–79; fan response to programming of, 80–84; multi-screen strategy of, 20, 58–60, 67, 86; as platform, 184; and quality TV branding, 132, 135; and Story Sync, 21, 61, 67–70, 74–75, 104
American Express, 93
American Idol (TV series), 42, 117, 124
amFAR, 171
Amplify (Twitter feature), 12
ancillary content, 41, 59–60, 86, 92, 180, 185. *See also* paratextuality
Anderson, Christopher, 132, 149, 151, 157
Andrejevic, Mark, 6, 62, 118, 124
Ang, Ien, 33
Annapurna Pictures, 138
appointment viewing, 26, 28, 101, 159, 179, 182
archive, 14, 16, 22, 62, 92
Aronczyk, Melissa, 62
Arrix, Kevin, 112
Arvidsson, Adam, 149
#AskScandal, 36–40, 42, 54
Associated Press, 36
AT&T, 89, 94, 104, 107
At Bat (Major League Baseball), 67
attention economy, 13–14, 119, 179
audio commentary, 30, 58, 63–68, 80, 83, 85. *See also* DVD
August, John, 122

aura: of brands, 62, 145, 149–50; of live television, 6–7, 33; of second-screen experiences, 21, 28, 55, 180
Auslander, Philip, 32
auteur: and Amazon Pilot Season, 134, 136, 139–41; authorship, 63, 69, 83; gender, 18, 32; and HBO, 123, 132–34, 150–53, 157–58, 162; quality TV discourses, 117, 132–34, 141; Shonda Rhimes as, 28–30, 34–42
authorship: and brands, 148, 151. *See also* auteur
automation, 19
Awl, The, 44

Bachelor, The (TV series), 113
badges: on check-ins, 21, 94, 101, 103–4, 107–8; on Foursquare, 93–94
Balderas, Leslie, 55
Ballers (TV series), 165
Ball in the Family (TV series), 177
Banet-Weiser, Sarah, 13, 145, 149–50, 171, 173
Banks, John, 125
Bans, Jenna, 34–35
Barbrook, Richard, 91
Battlestar Galactica (TV series), 30, 118
Becker, Ron, 170
Beers, Betsy, 30
Belson, Gavin (*Silicon Valley* character), 167
Ben and Kate (TV series), 100
Berland, Leslie, 176
Bessie (2015 film), 170
Best Buy, 89, 108–9
Betas (TV series), 124
Better Call Saul (TV series), 58–59, 70, 73, 80–81
Bewkes, Jeffrey, 158–59
Bezos, Jeff, 119, 140
Big Bang Theory, The (TV series), 102, 126
Big Little Lies (TV series), 80, 161, 163
Bjarkman, Kim, 66
Black, Shane, 139
Black audiences, 28, 31, 43–45
Black Lives Matter, 185
Black Twitter, 28, 44–45

Blatt, Ben, 25
blog post, 11, 14, 30, 58, 106
Bloomberg Media, 176
Blu-ray, 64
Boardwalk Empire (TV series), 85
Bochco, Steven, 134
Boddy, William, 8
Bogost, Ian, 95
Bolter, Jay, 8
Booth, Paul, 19, 91, 96
Bosch (TV series), 127–30, 133–34, 137, 141
Bourdieu, Pierre, 101
Bourdon, Jérôme, 197
Boyd, danah, 8
brand-creep, 22, 175
branding: and ABC, 26, 34, 47; and Amazon Studios, 123–25, 131, 136–40; and Black audiences, 31–32; and cable, 149–50; and check-ins, 93–95, 97, 102; consumer goods, 108; and digital media, 6, 13–14, 17, 65, 78; and holidays, 165–70; of platform authenticity, 159–65, 172–75; and masculinity, 18–19; and network second-screen experiences, 67, 113; and occupation of everyday life, 184–88; and Pride, 170–72; and quality TV, 117, 132–35, 140; and rebranding, 90, 120, 181; and Story Sync, 21, 77; and #TGIT, 45–47, 51–56. *See also* HBO
Breaking Bad (TV series), 67–70, 73, 81, 106
Brice, Wee-Bay (*The Wire* character), 161
broadcast television: and cable television, 8–9, 26, 134; and gender, 18, 60; and promotion of live television, 25–27, 32–36, 55; scheduling, 33–34, 47, 51, 55–56; and tension with Nielsen, 9–10. *See also* ABC
Brock, André, Jr., 18, 44–45
Brooker, Will, 74, 118, 154
Brookey, Robert, 63, 72
Browsers (TV series), 126
Bullock, Sandra, 157
Bush, George H. W., 37
buzz (social media), 4, 8, 32, 125, 131, 152
BuzzFeed, 160, 176–77

cable television: and broadcast television, 8–9, 26, 134; and cord-cutting, 11, 147, 159, 179; and gender, 18, 60; history of, 8–9, 14, 149; and on-demand film distribution, 64, 120; and original series production, 178–79; and quality TV, 123, 132–34, 149–50, 173. *See also* Story Sync
Café Society (TV series), 138
Caldwell, John T., 10, 17, 60, 64, 69–70, 76–78
Caluori, Sabrina, 85, 155
Cannes Film Festival, 138
Captain America: The First Avenger (2011 film), 100
Carah, Nicolas, 150, 169
Carpentier, Nico, 125
Carter, Chris, 127, 130, 134
Catch, The (TV series), 56
CBS, 11, 40, 67, 89, 120
CBS Interactive, 89
celebrity. *See* star
Chamberlin, Daniel, 111
Chandrasekhar, Jay, 135
channel: as distribution for repurposing, 62–65, 84, 153; for distribution of social media platform, 102, 125–26, 136, 157; as metaphor, 183–84
Chase, David, 134
chat room, 58, 75, 80, 85
check-ins: and badges, 21, 93–94, 101, 103–4, 107–8; cross-promotion of, 93–94, 101; development of, 21, 88–90; failure of, 90, 113–14; and fan capital, 92, 101–2, 114–15; features of, 90; Foursquare as example of, 21, 88, 93–94, 97, 103; and gamification, 90, 93–95, 103–4, 180, 182; investment in, 89–90, 93–94; and Letterboxd as example of, 115; and liveness of, 90, 92, 98, 100–104, 106–8, 110, 113; press coverage of, 97, 102–4, 107–8, 111–13; and ratings data, 13, 102; reward economy of, 22, 92–93, 113–15; rewards of, 21–22, 67, 145, 182, 187; social productivity of, 90, 92–93, 96–100, 102–8, 110–11, 113, 182; stickers of, 19, 21, 88–89, 93–94, 98–104,

113–14, 180; Telfie as example of, 115; TV Time as example of, 115. *See also* GetGlue; Miso; Viggle
chicken nuggets, 23
Chi-Raq (2015 film), 138
Christensen, Clayton, 190n13
chyron, 11–12, 26, 27, 33, 180
Cinco de Mayo, 165, 166
Clarke, Emilia, 144
cliffhanger, 34, 39–40, 42, 45–46, 55, 82, 174
Clorox, 89, 111–12
CNN, 68
co-connected viewing, 7, 27
co-creator, 58, 117, 151
code-switching, 44
Cohan, Lauren, 71–72
Cohen, Lara, 52
collective intelligence, 114, 124
color-blind casting, 30–31, 46
Comcast, 66, 89
Comedy Central Roast of Donald Trump (2011 TV special), 11–12
comic book, 66, 73, 86
commercials: for Amazon Studios, 116, 120–21; avoidance of, 9–10; and check-ins, 89, 94, 102, 108, 110–12, 114; in flow, 33; for *Off the Map*, 35; in promotional mix, 184; and Story Sync, 59, 78–79; for #TGIT, 46, 52–56; for #TGITWithdrawal, 53–56; for Viggle, 111. *See also* promotion
commodity activism, 171–72, 185
Community (TV series), 11
Complex (essay), 153
Condé Nast, 93, 177
connected viewing, 77
Connelly, Michael, 127–30, 134
Consiglio, Greg, 108
Consumer Electronics Association, 85
contests, 26, 33, 51, 58–60, 122
continuity, 33, 97
Cooper, Bradley, 3–5
Coppola, Roman, 127
cord-cutters, 11, 147, 159
cord-nevers, 147

Cosmopolitans, The (2014 TV pilot), 136
Couldry, Nick, 28, 32
Cranston, Bryan, 68
Creative Arts Emmys, 60
Crew, Amanda, 155
crisis historiography, 15–17
Crisis in Six Scenes (TV series), 139
cross-promotion: on check-ins, 93–94, 101; in HBO branding, 148–49, 153–54; in promotional strategy, 7; in #TGIT, 42, 46–47, 51; on Story Sync, 77–79, 84. *See also* promotion
crowdfunding, 117, 141–42
crowdsourcing, 104, 117, 122, 124, 142. *See also* fansourcing
Crunchyroll, 91–92
Cudlitz, Michael, 73
Cuse, Carlton, 30, 139

Daily Show, The (TV series), 126, 151–53
Dallas (TV series), 40
Dark Minions (2013 TV pilot), 126
Davis, Viola, 45, 48–49
Dawson, Max, 65–66, 166
Dawson's Creek (TV series), 74
Dayan, Daniel, 28
Deadline Hollywood, 182
Dead Run (mobile game), 78–79
de Certeau, Michel, 186
DeGeneres, Ellen, 3–5, 23–24
Deller, Ruth, 27
Demby, Gene, 44
Dexter (TV series), 105
DHARMA Initiative, 65–66
DiCaprio, Leonardo, 157
Dick, Philip K., 139
diegesis, 48, 76, 104, 163
Digi-Gratis economy, 91
digital enclosure, 6, 62, 78, 115, 181
digital video marketplace, 26, 64
Dijit, 90, 107
Dillon, Steve, 58
DirecTV, 89, 94, 104–5, 107, 110
DiRubba, Heather, 93
DISH Network, 105

Disney, 18, 158
dispersal, 64–66, 84. *See also* repurposing
disruption: as discourse and ethos, 7–10, 15, 89, 97, 121, 138, 142; historical examples of, 60; press coverage of, 42–43, 68, 94, 97, 112, 119–22
Divorce (TV series), 144–45
Dorsey, Jack, 176
Dove, 56, 114
Doyle, Gillian, 65
Dracula (TV series), 109
Dumenco, Simon, 113
Dunn, Jared (*Silicon Valley* character), 168
DVD: and advertisements, 78; and audience interpretations, 63, 71–72, 76; audio commentaries, 30, 63–68, 80, 83, 85; and authorship, 63, 69; behind-the-scenes feature, 60, 62–64, 71; collectors of, 63, 84; and interactivity, 64, 77, 85; and Netflix, 120; and repurposing, 64, 66, 69–72; and trivia, 64, 74
DVR: consumer adoption of, 26; industry anxiety over, 10–11; and Nielsen ratings, 10, 13; as on-demand technology, 33, 66; use of Viggle app with, 108, 110

Edge (2015 TV pilot), 139
Ehrmantraut, Mike (*Breaking Bad* character), 81, 105
electronic press kit, 64
Ellis, John, 32, 34, 47
Ellis, Kimberly C., 44
emojis, 25–27, 165, 169
empowerment, 20
engagement: as audience management strategy, 17, 20, 72, 80, 82, 86, 90, 93, 97–99, 104–8, 125, 142, 145, 151–54, 158–74, 186; as data, 8, 52–53; and ephemerality, 13–15, 182; and liveness, 11, 182; social media, 4, 11, 13, 22; and stars, 11, 21, 27–29, 32, 37–39, 42, 48–51, 144, 146, 150, 154–55, 157. *See also* platform authenticity
Ennis, Garth, 58
Enos, Mirielle, 56
Entertainment Weekly, 42, 101

enunciative productivity, 90, 95–96, 99
ephemeral historiography, 15–17, 20, 188–89
ephemeral labor, 118–20, 128–29, 141
ephemerality: and check-ins, 88–91; definition of, 14–16, 61; and labor, 118–20, 128–29, 141; and live television, 57, 60; and omnipresence, 146, 174, 185–87; snack content, 146, 166, 177; and social media, 24; and social productivity, 93, 104, 106, 113–14; and Story Sync, 66, 80, 84–86
Epix, 120
Eppink, Jason, 160–61
Epps, Jack, Jr., 121–22
ER (TV series), 33
Erigha, Maryann, 32
Evans, Elizabeth, 15
event: and liveness, 4, 11–12, 18, 25, 32–34, 42, 48, 54–55, 159–60, 170; and process of eventizing, 14, 23, 25, 26, 28, 32–34, 37, 118, 145, 153, 159–60, 168, 175, 187
Everett, Anna, 43
everyday life, 14, 22, 145, 148, 174–75, 186–87
Ewing, J. R. (*Dallas* character), 40
exchange economy, 91–93, 107–9, 112–14, 118
explicit productivity, 105

Facebook: and fandom, 11, 86; and GetGlue, 88–89, 92, 97–100, 102, 114; and HBO Connect, 155; and hype, 4; Live, 177; News Feed, 177–78; original programming of, 6, 13, 23, 176–80, 182; Places, 94; and pivot to video, 177; and Viggle, 108, 113, 114; Watch, 176–80
Face Off (TV series), 122
Fairey, Shepard, 40–41
Falling Skies (TV series), 105
fanboy, 18, 32, 58
fan capital, 92, 101–2, 114–15
Fandango, 108
fandom: and Amazon Pilot Season, 117–19, 121, 124–26, 130, 141–42; as default Social TV orientation, 16–20, 72, 180–81; and fanboy, 18, 32, 58; fan capital, 92, 101–2, 114–15; and fansourcing, 22, 117–19, 141–42,

181; and fansubbing, 91; and fan wiki, 65, 96; and gender, 18, 32, 58, 60; gift economy of, 91–92; and interactivity, 21, 59–60, 67–70, 75–77, 84, 85; and labor, 91–93, 118–19, 128–29, 141; and productivity, 90, 92–97, 99–100, 105–8, 110–11, 114, 182; on Reddit, 80–84; and repurposing, 64–66, 69–70, 84–86; reward economy of, 22, 92–93, 113–15; rewards for, 21–22, 67, 88–95, 98–113, 145, 182, 187; of Shonda Rhimes, 29–32; and trivia consumption, 60, 74, 96; on Twitter, 28, 37–38, 43–46, 51–56
fanfiction, 91
FanLib, 91–92
fan productivity, 90, 92, 95–96, 99, 114.
See also social productivity
fansourcing, 22, 117–19, 141–42, 181
fansubbing, 91
fan wiki, 65, 96
Fear the Walking Dead (TV series), 70, 75, 78–80, 83
feedback: in Amazon Pilot Season, 22, 121–30, 133, 135–43; culture of, 116–19, 142–43, 184; as interactivity, 77; on Reddit, 80–84
feedback culture, 116–19, 142–43, 184.
See also feedback
Feig, Paul, 157
Feuer, Jane, 32–33, 132
FIFA World Cup, 25, 110
Fincher, David, 123
Finke, Nikki, 122
Fiske, John, 90, 95–96, 99–101
fitness app, 94
Florini, Sarah, 18, 44
flow: definition of, 7, 27, 32–34; as digital movement, 78, 149, 172, 186; and DVD, 63; evolutions of, 34; of information, 17, 24, 64; and liveness, 7, 27–28, 32–34, 41–42, 46, 54, 182; and overflow, 74–75, 154; and scheduling, 36, 46–47, 51, 53, 66, 80; of second-screen experiences, 71–72; simulated, 21, 27–28, 41–42, 46, 51–57, 182, 184; tweet flow, 28, 42
Foley, Scott, 53

Food Network, 149
Foot, Kristen, 14
For Your Consideration campaign, 144–45, 167
Foursquare, 21, 88, 93–94, 97, 103
Fowler, Therese Anne, 139
Fox (TV network), 11, 31, 67, 100, 159
Foxx, Jamie, 99
Fring, Gus, 81
Fringe (TV series), 11, 106
Fuchs, Christian, 112
futurism, 9–10
FX (TV network), 132, 135

Galloway, Alexander, 19
Game Of Thrones (TV series): second-screen experiences for, 85; and Twitter, 144, 147, 154, 157, 159, 163–65, 168–69; Twitter TV ratings for, 42; #WinterIsHere campaign, 168–69
gamification, 62, 93–95, 104
Gap, 101
Gerlitz, Carolina, 97
GetGlue: and advertising, 102–3; and collective intelligence, 114; Conversation, 98–100; and corporate partnerships, 89, 94, 100–102; data, 102–3, 114, 181; and ephemerality, 88–91; Facebook, 88–89, 92, 97–100, 102; and fan capital, 92, 101–2; and fan productivity, 90, 92, 95–96, 99, 114; features of, 88–90, 92, 97–103; and gamification, 90, 95; and gift economy, 22, 90; and Guru system, 88–89, 99–101, 103, 180; interface of, 97, 100; investments in, 89; and liveness, 88–90, 92, 98–103; origins of, 88–89, 94; press coverage of, 89, 97, 102–3; rebranding of, 113–14; reviews on, 90, 99–101; reward economy of, 22, 92–93; rewards on, 88–89, 99–103, 108, 112–14; and social productivity, 21, 90, 92–93, 97–100, 114; and Sticker FAQ, 114; stickers of, 94, 98–103; as Social TV case study, 20–22; titles of, 19, 21, 88–89, 99–101, 113; user base of, 97, 102; and Viggle, 90, 112–13

GIF: and HBO, 144, 160–61, 163–65, 167, 168, 170, 172; as part of promotional mix, 53; and platform authenticity, 174, 184–85; and social productivity, 90, 96, 102
gift cards, 19, 88, 89, 94, 108–9, 111
gift economy, 22, 90–92, 113–14
Gilder, George, 9
Gillespie, Tarleton, 19, 184
Gillette, 110
Gilmore Girls (TV series), 140
Gitelman, Lisa, 8
Gladiators, 36–40
Glee (TV series), 11
Glee Project, The (TV series), 100
Gnip, 103
Goldberg, Evan, 58
Golden Globe Awards, 119, 136
Goldwyn, Tony, 25, 38–40, 47
Goliath (TV series), 140
Gone Girl (2014 film), 51–52
Good Girls Revolt (TV series), 139
Goodman, John, 126
Good Morning America (TV series), 32
Gossip Girl (TV series), 110
GoWalla, 94, 97
Grainge, Paul, 14, 61, 84, 153, 167
Grant, Fitzgerald (*Scandal* character), 25–26, 38–44, 46–47
Graver, Fred, 4
Gray, Herman, 31
Gray, James, 138
Gray, Jonathan, 16, 48, 62
Great Recession, The, 7
Green, Joshua, 65
Grey's Anatomy (TV series): and scheduling, 34–35, 37, 42; and Shonda Rhimes, 29–30; and #TGIT, 20, 25, 45–46, 48–51, 54, 57
Grimes, Rick (*The Walking Dead* character), 70, 72–73, 76
Grusin, Richard, 8, 183, 186
Gualtieri, Paulie "Walnuts" (*The Sopranos* character), 161

Hackett, Isa, 140
Hale, Tony, 167
Hand of God (TV series), 136
Hanks, Colin, 151–52
Hardwick, Chris, 59
Harrelson, Woody, 152–53
hashflags, 25–26
hashtag: as destination for conversation, 8, 25, 180, 185; #FYC, 144–45, 166; #Gladiators, 36–38; and HBO, 23, 144–47, 152, 154, 159–72; and platform authenticity, 174–76; #Pride, 170–72; #SiliconValleyHBO, 151, 155–57; #Veeple, 153–54. *See also* #AskScandal; #TGIT; #WhoIsQuinn; #WhoShotFitz; #WinterIsHere
Hastie, Amelie, 14, 61
Hastings, Reed, 123, 147
HBO: Amazon Channels, 137; and auteurism, 123, 132–34, 150–53, 157–58, 162; and binge-watching, 148, 161–65, 167; brand history of, 132, 148–50; and commodity activism, 170–72; competition with Netflix, 22, 123, 147–48, 158–59; Connect, 85, 154–55; documentary, 154; and everyday life, 173–75; and film library of, 146, 157–58; and Foursquare, 93; and *Game of Thrones*, 85, 144, 147, 154, 157, 159, 163–65, 168–69; and GetGlue, 94; Go, 85, 158–59, 162, 172; and golden age of television, 150; and holidays, 165–70; and Jeffrey Bewkes, 158–59; and *Last Week Tonight*, 144, 147, 152–53, 162; and Miso, 94, 105; "Not TV" slogan of, 18, 132, 135, 144, 150, 153, 158, 168, 173; Now, 147, 148, 159, 172; and omnipresence, 22, 146–47, 174; original programming of, 132–33, 135, 146, 149–50, 154–55, 157–59; and Pride, 170–72; and quality TV, 132–33, 135; and reruns, 146, 157; and Richard Plepler, 147; second-screen experience of, 61, 67, 85–86; and *Sex and the City*, 150, 169; and *Silicon Valley*, 7, 144, 151–52, 155–57, 160–62, 167–68, 170; as Social TV case study, 20, 22; and streaming, 148, 161–65, 167; and Sunday night, 97, 155, 157, 159, 161–62, 164, 168, 173, 175; and synergy, 152–53; and television criticism, 132–33,

146, 150, 152–53; and *The Sopranos*, 134, 150, 161, 165–66; universe of, 165–72; use of emoji by, 165, 169; use of GIF by, 144, 160–70; use of hashtag by, 144–45, 147, 152–57, 159–75; and user engagement, 154–56; and *Veep*, 144, 147, 153–54, 160–62, 166–68, 180; and #WinterIsHere campaign, 169–70; and work, 23, 165–67. *See also* platform authenticity
Hearst Magazines, 56
Hell on Wheels (TV series), 70
Helmond, Anne, 97
Hendricks, Richard (*Silicon Valley* character), 157, 161
Heroes (TV series), 65
High Maintenance (TV series), 165
Hills, Matt, 64
Hilmes, Michele, 134
Hilton, Shani O., 44
hipness, 101, 164
historiography, 15–17
Hoffmann, Gaby, 134
Hollywood Reporter, 42, 182
Holt, Jennifer, 17
Homeland (TV series), 106
home video, 11, 26, 64
House of Cards (TV series), 123, 132, 159
How to Get Away with Murder (TV series), 21, 25, 45
How to Make It in America (TV series), 93
Hulu, 11, 57, 116, 158, 178, 186
Humphreys, Sal, 125
Hurd, Gale Anne, 72
Huth, Denise, 72
hype: and ephemerality, 15, 120, 181; as paratext, 48, 62; Social TV, 6, 13–15, 181–82, 187; and trade press coverage, 42–43, 90, 103, 181–82
Hyundai, 89

Iampolski, Mikhail, 62
IFC Films, 138
I Love Dick (TV series), 140
immaterial labor, 118–19

immediacy: and live television, 4, 25, 27, 32–33, 40, 180–81, 184; and on-demand streaming, 118; and Twitter, 35, 38, 40, 113, 151
implicit productivity, 105
impressions (social media), 13, 27, 43, 55, 102
independent cinema, 138
Indiegogo, 117
Ingram-Waters, Mary, 56
In Living Color (TV series), 31
Instagram, 146, 150, 179, 185
interactivity: in branding, 149; the DVD, 63–64; and HBO second-screen experiences, 85; marketing of, 124–25; and participation, 17–19, 77–78; and remediation, 8–9, 11; and SHO Sync, 85; and Story Sync, 21, 59–60, 67–70, 75–77, 84; and Twitter, 145, 149, 157, 169; and Viggle, 108, 110, 113
interface: of Amazon Studios's website, 121, 125–29, 136; brands as, 149–50; of Facebook Watch, 178; of GetGlue, 97, 100; of Miso, 105; of/and platforms, 16, 19, 185; of Story Sync, 62, 72; of Viggle, 111
Internet Movie Database, 105
interstitial, 27, 33, 47
intertextuality: definition of, 59, 61–62; and HBO branding, 150; and Story Sync, 70–72, 74–75, 80
IntoNow, 67
Into the Badlands (TV series), 86
intratextuality, 72–73
iPad, 11, 120
iPhone, 11, 94
Iskold, Alex, 97–98, 101–2
iTunes, 11, 108, 109, 120
i.TV, 90, 103

Jacobs, Gillian, 151–52
Jean-Claude Van Johnson (TV series), 140
Jefferson, Thomas, 44
Jenkins, Henry, 17, 65, 69, 77
Johnson, Catherine, 148–49
Johnson, Derek, 60, 125
Johnson, Hassan, 161

Jolie, Angelina, 3–5
Joseph, Ralina, 31
Judge, Mike, 151

Kaling, Mindy, 151
Katz, Elihu, 28
Keating, Annalise (*How to Get Away with Murder* character), 50, 56
Kelley, David E., 139–40
Kellner, Jamie, 10
Kickstarter, 117
Killing, The (TV series), 70
Kindle Fire, 109, 120
King, Michael Patrick, 140
Kirkman, Robert, 83
Klein, Naomi, 148
Klinger, Barbara, 60–61, 63–64, 69, 71, 74
Kohan, Jenji, 159
Krinsky, Dave, 151
Kripke, Eric, 32
Kristeva, Julia, 62

Lagerwey, Jorie, 18, 60
Landecker, Amy, 135
Lannister, Jaime (*Game of Thrones* character), 165–66
Last Tycoon, The (TV series), 139
Last Week Tonight with John Oliver (TV series), 144
Late Night with Jimmy Fallon (TV series), 17, 110
Law, Jude, 163
Lawrence, Jennifer, 3–5
Lazzarato, Maurizio, 118
Lee, Hye Jin, 6, 62
Lee, Paul, 43
Lee, Spike, 138
Lefebvre, Henri, 174–75, 187
Leftovers, The (TV series), 162
legitimation: and awards, 187; process of, 18, 119–20; and quality TV, 131–36; and remediation, 7–9, 16, 179–82, 184
Letterboxd, 115
Leverage (TV series), 103

Levine, Elana, 18, 29–30, 33, 132, 134
Lewis, Joe, 135, 139
Lexus, 89
LG, 85
Lifetime, 149
Likert scale, 128
likes, 3, 8, 14, 19, 97–101, 180
Lincoln, Andrew, 70
Lindelof, Damon, 30
liveness: and advertising, 9–12, 78–80, 110–12, 114; and appointment viewing, 26, 28, 101, 159, 174, 179, 182; and check-ins, 98–112; construction of, 27, 33–34; definition of, 7, 27, 32–34; and events, 3–5, 25, 28, 139, 177, 179; flow, 7, 27–28, 32–34, 41–42, 46, 54, 182; and immediacy, 4, 25, 27, 33, 35, 38, 40, 180–81; scheduling, 36, 46–47, 51, 53, 66, 80; of second-screen experience, 66–68, 75–80, 84, 86; simulated, 21, 27–28, 40–42, 46, 51–57, 182, 184. *See also* live-tweeting
live-tweeting: and data, 25, 42, 52; early experiments with, 11–12, 27; and promotion by Twitter, 10–11; and *Scandal* cast, 21, 26, 36–45, 181; as standard Social TV practice, 11–12, 53, 86, 88, 118, 145; and #TGIT, 45–55, 57
Living Single (TV series), 31
location-based service, 97
London, Michael, 124
Lonergan, Kenneth, 138
Lord of The Rings, The (franchise), 118, 139
Los Angeles, 128
Los Angeles Times, 46
Lost (TV series), 30, 65, 118, 139
Lost City Of Z, The (2016 film), 138
Lost Experience, The (ARG), 65–66, 78
Louis-Dreyfus, Julia, 167
Lovink, Geert, 96
Lowes, Katie, 39, 48–49
Low Winter Sun (TV series), 70
loyalty program, 94–95, 107
Lury, Celia, 149–50, 157
Lynch, Jane, 136

Mackenzie, Madeline (*Big Little Lies* character), 161, 163
Mad Men (TV series), 67, 102, 139
Magnolia Pictures, 138
Major League Baseball, 67, 176
Major League Soccer, 18
Manchester by the Sea (2016 film), 138
Man in the High Castle (TV series), 139–40
Manjoo, Farhad, 44
Manovich, Lev, 19, 186
Marvel Studios, 18, 148
Marvelous Ms. Maisel (TV series), 140
masculinity, 18, 60, 133, 141
Mazin, Craig, 122
McCarthy, Anna, 186
McCarthy, Melissa, 157
McConaughey, Matthew, 152–53
McDonald, Paul, 63–64
McDonald's, 89, 109
McGorry, Matt, 53
McKean, Mac, 60, 67–69, 78–79
McKidd, Kevin, 50
McPherson, Stephen, 30
meme: and award shows, 4–5; and ephemeral media, 22–23, 55; and platform authenticity, 13, 26, 144, 146, 148, 160–72, 185; and social productivity, 96, 114
message board, 96
Meyer, Selina, 166–68
micro-moment, 15, 187
Middleditch, Thomas, 161
Midha, Anjali, 47
Mildred Pierce (2011 TV miniseries), 162
millennial, 18, 171
Miller, Vincent, 96
Minnelli, Liza, 3
Miramax Pictures, 138
Miso: and Andrew Seroff, 93, 105–7; and advertising, 112; badges of, 89–90, 94, 103; corporate partnerships, 89, 94, 104, 106; and crowdsourcing, 104, 106; and data, 113; and Dijit, 90, 107; and ephemerality, 88–91, 106; and explicit productivity, 105; and gamification, 90, 105; and gift economy, 22, 90; and Facebook, 92; and *Falling Skies*, 105; Fan Club of, 103–4; fan productivity, 90, 92, 103–7; features of, 88–90, 103–5; and implicit productivity, 105; liveness, 88–90, 104–5, 107; and NextGuide, 90; origins of, 88–89; and point system of, 88–90; purchase of, 90, 107, 113; Quips of, 105–7; reward economy of, 22, 89–90, 92, 103–4, 113; rewards on, 90, 103, 107–8; SideShows of, 104; and social productivity, 21, 90, 103–7, 113; as Social TV case study, 20–22; Somrat Niyogi, 104; and trivia, 104; user base of, 90, 105
mixed economy, 91–92
Mixed Match Challenge (TV series), 177
mobile application, 4, 6, 14, 19. *See also* check-ins; second screen; Story Sync; Twitter
mobisode, 10–11
Modern Family (TV series), 111
Moore, Alison, 67
Moore, Ronald, 30
Mo.Pho.to, 94
Mozart in the Jungle (TV series), 127, 130
Mukherjee, Roopali, 171
multiculturalism, 28, 31
multi-platform: and buzz, 32; ecosystem of, 23; and omnipresence, 174, 187; and promotional material, 20, 21, 23, 26, 28, 35, 41, 45–46, 49, 174; and repurposing, 61, 65, 72, 74, 157
multi-screen: and data, 6, 47; and digital enclosure, 6, 62, 78, 115, 181; and habits, 11, 47, 66; and industry strategy, 13, 19, 27–28, 67, 125, 145, 181; and navigation, 16, 28, 32, 64, 75–76; simulated liveness and flow, 27–28, 36, 39–42, 46–55; and synchronized reiteration, 60, 68, 70–78; and viewing, 4, 6, 13, 19, 27–28, 66–67, 162. *See also* check-ins; second screen; Story Sync
multitasking: sensibility of, 61, 187; and viewing, 7, 11, 19, 69, 86, 96, 98, 110, 146, 187
Murray, Simone, 17, 118
Must-See TV, 47

narrative complexity, 132, 150
National Association of Television Program Executives, 85
National Donut Day, 165
National Geographic, 94
National Public Radio, 44
NBCUniversal, 158
NCAA men's basketball tournament, 110
Negan (*The Walking Dead* character), 71–72, 82
Negroponte, Nicholas, 9
Netflix: and competition with Amazon, 135–36; and competition with HBO, 22, 123, 147–48, 158–59; and Nielsen ratings, 182; original programming investments of, 13, 57, 116, 123, 132, 178–79, 183; and recommendations, 186; and Shonda Rhimes, 56–57; as streaming video portal, 11, 20, 64, 120, 135
network: as metaphor, 183–84. *See also* broadcast television; cable television
New Girl (TV series), 99–100
New Line Cinema, 118
Newman, Michael Z., 18, 30, 132, 134, 160
News Corp, 158
New York Film Festival, 138
New York Times, 30, 93, 123
New Yorker Presents, The (TV series), 139
NextGuide, 90
Nickelodeon, 149
Nicotero, Greg, 71
Nieborg, David, 104
Nielsen: DVR ratings, 10, 13; live television ratings, 26, 39, 58, 119, 184; and tension with networks, 8–10; Twitter TV ratings, 12–13, 25, 42–43, 47, 51–53, 57, 102, 181–82
Night Of, The (2016 TV miniseries), 162
Niyogi, Somrat, 104–5
Nowalk, Peter, 47
Nygaard, Taylor, 18, 60
Nyong'o, Lupita, 3–5

Obama, Barack, 3, 41
Oscar selfie, the, 3–5, 13, 23
Office, The (US TV series), 151

Off the Map (TV series), 34–36
Oliver, John, 144, 151–53
Olympics, 67
omnipresence, 22, 146–47, 174, 184–87
Once Upon a Time (TV series), 109
on-demand media: and digital technologies, 6–7, 11, 34; and set-top boxes, 64; and streaming video portals, 22, 118, 125, 158, 179–80; and viewing habits, 86, 88, 146, 179–80, 184
One Mississippi (TV series), 140
On the Lot (TV series), 122
Orange Is the New Black (TV series), 159
Ott, Brian L., 170
Ouellette, Laurie, 171–72
overflow, 74, 154
Overmyer, Eric, 127
Oxygen, 101

Page, Elliot, 136
Papacharissi, Zizi, 44
paratextuality: and authorship, 68–69, 148; definitions of, 16, 61–62, 154; fan response to, 83–84; and repurposing, 63–66, 70–80
Parker, Deborah, 63, 71
Parker, Mark, 63, 71
Parker, Sarah Jessica, 144
participation: and Alternative Reality Game, 11; Amazon Studios, 22, 117, 119, 124, 127–29, 133, 140–42; and crowd-sourcing, 117; and interactivity, 77, 86; and labor, 125; as promotional discourse, 8, 22, 51, 72, 84, 91, 113–14, 117, 119, 125, 127, 140–42, 169, 181, 185–86; and Social TV environment, 4–6, 16, 59, 79–80, 119, 169, 180, 185, 189
Paskin, Willa, 133
Paul, Michael, 137
PBS, 94, 137
performative allyship, 172
Perk, Inc., 90
Perkins, Quinn (*Scandal* character), 39–40
Perlman, Ron, 136
Perren, Alisa, 17, 138

phatic communication, 19, 96
pilot episode, 22, 36–37, 58, 73, 181
Pitt, Brad, 3–5
pivot to video, 177
Pizzolatto, Nic, 152–53
place-shifting, 11, 60, 78, 159
platform: in definition of Social TV, 6–7; and omnipresence, 22, 146–47, 174, 184–85; television as, 180–89; YouTube as example of, 178–79. *See also* check-ins; Facebook; interactivity; multi-platform; platform authenticity; Story Sync; Twitter
platform authenticity: and authenticity, 145–46; and binge-watching, 161–65; and commodity activism, 170–72; definition of, 22, 145–48; and everyday life, 173–75, 185; and GIF, 159–61; and hashtag, 159–74; and HBO, 22, 146–48, 158–75; and holidays, 165–70; and omnipresence, 146–47, 174, 184–85; and Pride, 170–72; and specificity, 146; and streaming, 161–65; and work, 23, 165–67
Plec, Julie, 17
Plepler, Richard, 147
podcasts, 10, 30, 65, 185
Polan, Dana, 150, 153, 164
Pompeo, Ellen, 46, 48, 50
Pope, Olivia, 25–26, 37–40, 43–44, 51, 53, 56
post-racism, 30–32
Preacher (comic series), 58
Preacher (TV series), 58, 86
Price, Roy: and Amazon Studios's founding, 116, 120; and Amazon Studios's rebranding, 138–40, 142; and Pilot Season feedback, 122–27, 129–30; and quality TV discourse, 133, 135
Pride month, 170–72, 185
Primetime Emmy Awards, 145
Private Practice (TV series), 34–35
procrastination economy, 13–14, 96, 146
product placement, 4, 7, 9, 70, 78
produsage, 117
produser, 117, 125
Project Greenlight (TV series), 122
Property Brothers, The (TV series), 79

promotion: and affective labor, 92; for Amazon Pilot Season, 117–18, 120, 123, 131–32; and cross-promotion, 7, 42, 46–47, 51, 77–79, 93–94, 101, 148–49, 153–54; and DVD, 63–64, 72; and Facebook Watch, 178; and flow, 33; and For Your Consideration, 144–45, 167; on Foursquare, 93–94; and hashflags, 25; for HBO, 18, 93, 144–55, 157, 158, 164, 166, 169–72, 181; key art as, 110; and liveness, 33, 100; for *Off the Map*, 34–36; and premediation, 186; and remediation 8; and repurposing, 59–60, 77–80, 84; for Samsung, 4; for *Scandal*, 38–44; and second-screen experiences, 67; and showrunners, 29; as site of analysis, 6, 16, 19; and social productivity, 102–3; on Story Sync, 77–80; and stickers, 89, 98–99; for #TGIT, 21, 26–28, 45–56, 181–82; for TiVo, 10; for #TweetWeek, 11; on Viggle, 109–10
Provencio, Marla, 52
Psych (TV series), 67
publicity: authorship, 148; and Carter Wilkerson, 23; and feedback, 118, 122; and For Your Consideration, 145; for HBO, 145, 148, 150, 152, 155; and interviews, 65, 149, 152; for Kerry Washington, 36; for Miso, 106; and paratext, 148; and photography, 56, 150; and Pride, 170; as site of analysis, 16, 21–22, 92; for Story Sync, 68, 77; and television criticism, 152; for #TGIT, 21, 26–28. *See also* promotion
Purcell, Gavin, 17

quality TV: as branding strategy, 132–36, 149–50, 173; challenges of, 137–41; definition of, 117, 132; and legitimation, 18, 22; press celebration of, 119, 132–33, 136, 141
Queen Latifah, 170
Quips, 105–7

Really (2014 TV pilot), 135–37
Rebels, The (2013 TV pilot), 130–31
Redbox, 64
Reddit, 62, 80–84, 86

Red Oaks (TV series), 136–37
red wine, 45, 50, 56
remediation: definition of, 7–8; historical examples of, 9–10; as industry discourse, 23, 107, 113; and social platforms, 177–79, 182–83, 186–87
remediation-legitimation cycle: definition of, 7–9; and ephemeral historiography, 15–17; evolution of, 179–81; press coverage of, 182
remote control, 9, 34
ReplayTV, 10
representation: and race, 28, 30–32, 37, 43–44; and sexuality, 136, 170–72
repurposing: definition of, 60; as digital strategy, 64–66, 69–70, 84–86, 184; and multiplexing, 157; and paratexts, 21, 62, 64–66; and Story Sync, 70–79
rerun, 11, 146, 157, 178
retweet: and Black Twitter, 44; and Carter Wilkerson, 23–24; from Amazon Studios, 136; from HBO, 144, 147, 151–57; as interactive feature, 19; from *Scandal* cast, 36–38, 40, 49; from Shonda Rhimes, 35; of Oscar selfie, 3; as promotional strategy, 180, 184
Revenge (TV series), 99
reward economy, 22, 92–93, 113–15
rewards: as check-in strategy, 21–22, 67, 145, 182, 187; economy of, 22, 92–93, 113–15; on GetGlue, 88–95, 98–103; on Miso, 103–7; on Viggle, 108–13
Rhimes, Shonda: departure from ABC, 56–57; and race, 31–32, 43–45; as social auteur, 28–30, 34–42; and #TGIT, 20–27, 46–55
Riffaterre, Michael, 62
ritual, 28, 174
Roadside Attractions, 138
Rogen, Seth, 58
Ross, Matt, 167
Royal Caribbean, 89, 109

Salamanca, Hector, 81
Samsung, 3

San Diego Comic-Con, 18
Sandvoss, Cornell, 96
Santo, Avi, 150–51, 157, 168
Saturday Night Live (TV series), 98
Saunders, George, 140
Scandal (TV series): and Black Twitter, 44–45; and check-ins, 100; and live-tweeting, 34–39, 43, 48–51, 180–81; and #TGIT, 20, 25–28, 45–47, 55–57; and #TGITWithdrawal, 52–54; and #WhoIsQuinn, 39–40; and #WhoShotFitz, 40–42, 46–47
Scannell, Paddy, 174
Schäfer, Mirko, 105
scheduling, 47–48, 54, 56, 62
Schneider, Steven, 14
Schwartzman, Jason, 127
Sconce, Jeffrey, 65, 73
Scott, Drew, 79
Scott, Suzanne, 18, 60, 91, 96
screenshot, 14, 80–84, 105, 107
second screen: devices as, 6–7, 11, 180; and mobile applications, 14, 16, 19, 180–81, 187; and simulated liveness and flow, 27–28, 47–49; and #TGIT, 48–52; usage of, 12–13. *See also* check-ins; live-tweeting; Story Sync; synchronized reiteration
semiotic productivity, 90, 95–96, 99
Sepinwall, Alan, 152
Seroff, Andrew, 93, 105–7, 110, 112, 114
Sex and the City (TV series), 140, 150, 165, 169
Shazam (mobile app), 94
Sherman-Palladino, Amy, 140
Shondaland, 21, 27, 45–55
Short, Columbus, 36–37
SHO Sync, 85–86
showrunner: as auteur, 65, 68, 134, 139; and celebrity, 30–32; as fanboy, 32, 58; and fan engagement, 19, 29–30, 32
Showtime: as Amazon Channels partner, 137; and cable brand identity, 132, 135, 159; as check-in partner, 89, 94, 105; and second-screen experiences, 61, 85–86
Sicha, Choire, 44
SideShows, 104–7
signifyin', 44

Silicon Valley: and disruption discourse, 7, 20, 85, 89, 97; and fandom, 18–20, 95–96; and mergers and acquisitions, 90, 94; and partnerships with Hollywood, 11–12, 21, 102–3, 119, 187–88; and remediation-legitimation, 8–10, 180–82; and Social TV hype, 5–6; and start-up failure, 15–17, 107

Silicon Valley (TV series), 7, 144, 151–52, 155–57, 160–62, 167–68, 180

Silverstone, Roger, 174

Simpsons, The (TV series), 4, 31

simulated liveness and flow: definition of, 21, 27–28; and *Scandal*, 36, 39–42; and #TGIT, 46–55

Sirico, Tony, 161

Six Feet Under (TV series), 127

SKAM Austin (TV series), 177

Slate, 42, 44, 97, 131, 133

smartphone, 3–6, 11–12, 50, 85, 104, 111

Smith, Greg M., 64

Smith, Jo T., 77

Smith, Judy, 37

Smith, Kevin, 138

Smith-Shomade, Beretta E., 31

snacking, 52, 146, 163–64, 166, 174, 185

social impressions, 27, 55, 102

social media. *See* social platform

social platform: and algorithm, 19–20; and auteur, 32; and data, 114; and fandom, 17–19, 103, 112–13; and integration with second screen, 57, 89–90, 93; and meme, 13; and overflow, 154; and participation, 142; and simulated liveness and flow, 21, 27–28; as site of analysis, 16–17, 20, 24; in Social TV mix, 6–7; and television-as-platform, 180, 185. *See also* ABC; Facebook; HBO; platform authenticity; Social TV; Twitter

social productivity: definition of, 90, 92–93, 96; examples of, 97–100, 105–8, 110–11, 182; as promotional strategy, 102–3, 105–8, 110–11, 113

Social TV: definition of, 6; and ephemerality, 14–16, 23; and everyday life, 14, 22, 174–75, 186–87; failure of, 14–17, 20–21,

187–88; and fandom as default orientation, 16–20; hype of, 6, 13–15, 181–82, 187; media conglomerate investment in, 66–67, 93–94, 101–2, 105, 112, 176–77; metrics of, 12–13, 27, 51–53, 102, 181–82; origins of, 4–11, 13; press coverage of, 4–8, 13, 16, 181–82, 187; and remediation, 6–10, 177–80; and social media and venture capital, 4, 91–92, 103. *See also* Amazon Pilot Season; check-ins; ephemeral historiography; ephemeral labor; fansourcing; HBO; platform; remediation-legitimation; *Scandal*; social productivity; Story Sync

Soderbergh, Steven, 136–38

Soloway, Joey, 124, 127, 130, 134–36

Sons of Anarchy (TV series), 134

Sony, 123

Soprano, Tony (*The Sopranos* character), 165–66

Sopranos, The (TV series), 134, 150, 161, 165–66

Soundcloud, 114

South by Southwest, 44, 94

Spigel, Lynn, 166

Sports Illustrated, 177

Stanfill, Mel, 18, 129

star: in behind-the-scenes feature, 71, 74; in cross-promotion, 78–80; as iconography, 89; in quality TV discourse, 126, 131, 136, 146, 150, 153, 158, 173; on Twitter, 11, 21, 27–29, 32, 37–39, 42, 48–51, 144, 146, 150, 154–55, 157

Starbucks, 89, 108

Starz, 137

Stein, Louisa, 18

Stephenson, Chris, 95, 107–8, 110–11

Sticker FAQ, 114

stickers: on Foursquare, 93–94; on GetGlue, 94, 98–103; on Miso, 104; as rewards, 19, 21, 88–89, 113–14, 180

Stillman, Whit, 136

Stone, Biz, 11

Story Sync: Before & After feature of, 73; and *Better Call Saul*, 59, 70, 73–74; and

INDEX

Breaking Bad, 68–70; and cross-promotion, 78–79; development of, 21, 58–59, 60–61, 67–68; as digital enclosure, 62, 75, 78, 180; end of, 84–87; and ephemerality, 61–62; fan response to, 80–84; and *Fear the Walking Dead*, 75–76, 78–79; Gore Gauge feature of, 68, 75, 77; Graphic Origins feature of, 72–73; Judgment feature of, 68, 76–77, 79, 81–82, 110; Mastermind Meter feature of, 68; Photo Flashback feature of, 68; promotion for, 67–70; repurposing, 70–79; and similarities to check-ins, 104, 110; and storyworlds, 58–60, 65, 71–75, 77; and synchronized reiteration, 60, 68, 70–78; Tactical & Morality Matrix feature of, 75–76; Thematic Callback feature of, 68; and *The Walking Dead*, 21, 59, 61, 67–77, 180
storyworld: of *Better Call Saul*, 59, 70, 73; of *Breaking Bad*, 68–70, 72; digital ephemera, 58; of *Fear the Walking Dead*, 75–76, 78–79; Story Sync's construction of, 58–60, 65, 71–75, 77; of *The Walking Dead*, 59, 70–77; and transmedia storytelling, 65–66
Straw, Will, 105
streaming video portal, 26, 64, 158, 178–83, 187. *See also* Facebook Watch; HBO; Hulu; Netflix
Streep, Meryl, 3–5
Striphas, Ted, 129
subcultural capital, 92, 101–2
subculture, 92, 101–2
subscription business model, 137, 141–43, 146–47, 157, 159, 185
Subway, 89
Sundance Film Festival, 138
Super Bowl, 11, 25
Supernatural (TV series), 32, 58
Survivor (TV series), 11
Sutter, Kurt, 134
Swan, Phillip, 9
sweeps (ratings), 37
synchronized reiteration: definition of, 21, 60; examples of on Story Sync, 68, 70–78
synergy, 41, 153, 168

tab (browser), 78
Talking Dead (TV series), 59, 79
Tambor, Jeffrey, 127, 134
Tarantino, Quentin, 138
Targaryen, Daenerys (*Game of Thrones* character), 144
Target, 89
Tartikoff, Brandon, 124
TechCrunch, 7, 107
tech industry. *See* Silicon Valley
technological utopianism, 7–9
television criticism, 29, 122, 132–37, 152–53
Television Critics Association, 122, 134–35
television screen: as access point for streaming portal, 178; as part of second-screen experience, 7, 16, 27, 47, 69–70, 78
Telfie, 115
temporality, 61
test screening, 117, 121–22
textual productivity, 90, 95–96, 99
TGIF, 25–26, 47
#TGIT: early campaign, 45–47; end of, 56–57; live-tweeting, 48–50, 181; success of, 21, 25–28, 54–55, 182; and #TGIT-Crossover, 56; Tweepstakes, 51–52
#TGITWithdrawal, 53–54, 56
30 Rock (TV series), 33, 126
Thomas, June, 97
Thompson, Robert J., 132
Thornton, Sarah, 101
Those Who Can't (2013 TV pilot), 126–27
Thursday Night Football, 139, 177
TicToc (streaming news service), 176
tie-in novel, 66
TikTok, 176, 179
time-shifting, 9, 28, 40, 172
Time Warner, 66–67, 89, 158–59
TiVo, 10. *See also* DVR
TNT, 103
Today (TV series), 24
Tomlinson, John, 65
trailer, 33–34, 37, 41, 45–46, 62, 168
transmedia storytelling: and authorship, 65–66; and canon, 72; and commercials,

78; as repurposing, 60–62, 69; strategies of, 10, 149
Transparent (TV series), 119, 124, 127–28, 130, 133–37
Trendrr, 12
trivia: contests on Story Sync, 58–60, 67–68, 74–75, 84–85; as repurposing, 60–62, 69; as social productivity, 96; on SHO Sync, 85; on Viggle Live, 110, 113
Trudeau, Garry, 126
True Blood (TV series), 85, 97, 165
True Detective (TV series), 152–53, 159
Trump, Donald, 11–12, 102
Tumblr, 86, 160
Turn: Washington's Spies (TV series), 70
Turner, Graeme, 106
Tussey, Ethan, 174, 185
Tvtag, 90, 113
TV Time, 115
tweet-peat, 11
Tweet Week (promotional event), 11
Twentieth Century Fox, 51, 123
Twitter: Academy Awards, 3–5; activism on, 22, 146, 170–74; Amplify, 12; and Black Twitter, 28, 44–45; branding, 13, 150; and check-ins, 89, 91, 97–99, 102, 114; *Comedy Central Roast of Donald Trump*, 11; and commodity activism, 170–72; conversation on, 25, 27–28, 32, 36, 38–39, 41–45, 47, 51, 55–56, 58; cross-promotion on, 42, 46–47, 52, 148–49, 153–54; and data, 3, 12–13, 42–43, 51–53; early television experiments with, 11–13, 17–18; and emoji, 25–27, 165, 169; evolution of, 145; and GIF, 53, 160–65, 167–68, 170, 172, 174; impact on television viewing of, 12; and links, 14, 48, 151–52, 170; meme on, 4, 13, 26, 55, 146, 159, 178; replies, 23, 29, 30, 36–37, 55, 117, 147, 161, 169; Shonda Rhimes activity on, 29–30, 34–35, 38, 50; and simulated liveness and flow, 21, 27–28, 36, 39–42, 46–55; as site for analysis, 6, 19, 21–22, 147; and Story Sync, 59, 61–62, 67, 86; as television partner, 11–12; trending topics on, 25, 42, 55, 145, 147, 155, 185; tweet-peat, 11; and

Tweet Week, 11. *See also* hashtags; HBO; live-tweeting; platform authenticity; retweet; *Scandal*; #TGIT, early campaign; Twitter TV Ratings
Twitter TV Ratings: creation of, 12–13; and GetGlue ratings reports, 102; press coverage of, 12–13, 47, 51–53, 181–82; success of #TGIT series within, 25, 42–43, 51–53, 57. *See also* Nielsen

Uber, 7
universe: HBO, 165–72; of the Marvel Cinematic Universe, 48; of #TGIT/Shondaland, 21, 27, 45–55; of *The Walking Dead*, 70–76; and transmedia storytelling, 65–66
upfront, 43, 176
UPN, 31
Uricchio, William, 34, 183
USA Network, 67
USA Today, 94
user-generated content, 91–93, 98–101, 103–7, 177, 183

Vampire Diaries, The (TV series), 17, 98
van Dijck, José, 19, 28, 42, 104, 108
Vanity Fair, 177
Variety, 29, 37, 123, 131, 136, 138, 140
VCR, 9, 33
Veep (TV series): promotion on Twitter of, 144, 147, 153–54, 157, 180; use of GIFs in promotion of, 160–62, 165–68
Venkataraman, Prakash, 107
venture capital, 4, 91, 103
verisimilitude, 74
Verizon, 66–67
Vernoff, Krista, 30
Viacom, 66–67
Viggle: and advertising, 108–13; and Andrew Seroff, 93, 110, 112–13; Audience Network, 112; and Chris Stephenson, 95, 107; and collective intelligence, 114; and consumer goods, 89, 109–11; and corporate partnerships, 89, 108–10, 112–13; and data, 111, 113–14; digital fingerprint technology of, 89, 94, 114; and ephemerality, 88–91;

and exchange economy, 109–12; and gamification, 90, 95, 110; and GetGlue, 90, 112–13; and gift cards, 89, 108–9, 111; and gift economy, 22, 90; features of, 88–90, 108–9; and Greg Consiglio, 108, 112; and interactivity, 108, 110, 112; interface of, 109–10; and Kevin Arrix, 112; Live, 110; and liveness, 88–90, 109–11; loyalty program of, 94–95, 107–8, 113; origins of, 88–89; and personalization, 110–11; and point system of, 88–90, 108–14; press coverage of, 108, 112; promotion of, 111–12; and purchase by Perk, Inc., 90, 113; reward economy of, 22, 92–93, 111–14; rewards on, 88–90, 108–14; and Shazam, 94; and social productivity, 21, 90, 111–14; as Social TV case study, 20–22; and Soundcloud, 114; and *The Bachelor*, 112–13; and trivia, 110; user base of, 90, 113; and Vigglers Twitter account, 114
virtual reality, 61, 86
Vox Media, 176
Vulture, 152–53

Walking Dead, The (comic series), 59, 71–73
Walking Dead, The (TV series): check-in activity for, 98, 106, 109; digital ephemera related to, 86; fan response to, 82–84; and Story Sync, 21, 59, 61, 67–77, 180; and storyworld of, 70–77
Walsh, Gary (*Veep* character), 167
Walsh, Matt, 153
Wanzo, Rebecca, 18
Ward, Sam, 149
Warner, Kristen, 18, 31, 43
Warner Bros., 102, 121
WarnerMedia, 158
Washington, Kerry: as Olivia Pope (*Scandal* character), 25, 43, 56; and Twitter, 27, 36–40, 42, 49–50, 54–55, 181
watchlist, 168, 177–78
WB, The, 31
Weather Channel, 149
Weiner, Matthew, 139
Weinstein, Harvey, 138, 140

Weinstein Company, 138
Weitz, Paul, 127
Wendy's, 23–24
Westerfelhaus, Robert, 63
Westworld (TV series), 80, 164–65, 172
Wheatley, Helen, 15
White, Walter (*Breaking Bad* character), 68, 81
#WhoIsQuinn, 39–41
#WhoShotFitz, 40–42, 46–47
#WinterIsHere, 168–69
Wilkerson, Carter, 23–24
Williams, Maisie, 154
Williams, Raymond, 7, 33, 173
Williamson, Kevin, 17
Wire, The (TV series), 127, 150, 160–61, 169–70
Witherspoon, Reese, 161, 163
Women's National Basketball Association, 176
Wonder Wheel (2017 film), 138
Woods, Zach, 168
word of mouth, 102
workplace, 94, 145, 166–68
workweek, 23, 29, 167
world-building, 73
World Health Organization, 83
writers' room, 30, 69

X Factor, The (TV series), 67
X-Files, The (TV series), 127, 134

Yahoo!, 67, 91
Young Pope, The (TV series), 163
YouTube, 13, 178–79
YouTube TV, 183

Zagat, 93
Zeebox, 66
Zenith, 9
Zombieland (TV series), 126
zombies, 67–68, 71, 73, 76, 126
Zook, Kristal Brent, 31
Z: The Beginning of Everything (TV series), 139

ABOUT THE AUTHOR

Photo courtesy of the author

Cory Barker is assistant professor of communication at Bradley University in Peoria, Illinois, where he teaches classes on journalism, digital media, and Hollywood. His research focuses on the global media industries' efforts to overcome and harness the ephemerality of the modern attention economy. He is coeditor of three books, including *The Age of Netflix: Critical Essays on Streaming Media, Digital Delivery, and Instant Access*, and his work has been published in *From Networks to Netflix: A Guide to Changing Channels*, *#WWE: Professional Wrestling in the Digital Age*, *Television & New Media*, and *Transformative Works and Cultures*. As a critic and cultural commentator, his work has appeared in *The A.V. Club*, *Complex*, *Vox*, TV.com, and *TV Guide*.

www.ingramcontent.com/pod-product-compliance
Lightning Source LLC
Chambersburg PA
CBHW030615230426
43661CB00053B/1992